LOOKING FOR LEARNING: AUDITORY, VISUAL AND OPTOMOTOR PROCESSING OF CHILDREN WITH LEARNING PROBLEMS

BURKHART FISCHER

Nova Science Publishers, Inc.
" ª© «±ℬ

For permission to use material from this book please contact us:
Telephone 631-231-7269; Fax 631-231-8175
Web Site: http://www.novapublishers.com

NOTICE TO THE READER

Library of Congress Cataloging-in-Publication Data
Fischer, Burkhart.
  Looking for learning : auditory, visual, and optomotor processing of children with learning problems / Burkhart Fischer.   p. ; cm.
  Includes bibliographical references and index.
  ISBN-13: 978-1-60021-502-5 (hardcover)
  ISBN-10: 1-60021-502-5 (hardcover)
 1. Learning disabilities–Diagnosis. 2. Learning disabilities–Treatment. I. Title.
DNLM: 1. Learning Disorders–diagnosis. 2. Learning Disorders–therapy. 3. Auditory Perception–physiology. 4. Child. 5. Eye Movements–physiology. 6. Visual Perception–physiology. WS 110 F529L 2006
RJ496.L4F57   2006
618.92'85889–dc22   2006039136

$- ´œ¥ \quad ^{a\,o} \ a\!\beta \ `` \pm^{TM}\!\!\!V\!\!\!/ \ \tilde{O} \ ^a \quad ^a -´œ¥ \quad ^a\!ÆÙ \quad Ú\!❖ \ `` ^a© \ll\pm\!Æ\!B$

LOOKING FOR LEARNING: AUDITORY, VISUAL AND OPTOMOTOR PROCESSING OF CHILDREN WITH LEARNING PROBLEMS

Contents

References   175

Index   181

Preface

This book is written at a time, when the requirements for successful school careers are very high: children have to learn many more facts and cover many different fields of science, e.g. Biology, Chemistry, Physics. We may think that this is only a question of memory, dilligence and especially intelligence. It therefore comes as a surprise, when we meet children at school who have considerable problems in learning. They have spent their early years in quite the same way as others without any signs of illnesses or abnormalities. Yet, they begin to struggle at school, when they try to learn to read or write or to use numbers. Attempts to blame low intellectual skills for the failure fail, because the children do well as soon as reading or writing is not needed. Today the phenomena of dyslexia and dyscalculia are well known, but the problem of remediation has not yet been solved. Many attempts are made to find the causes for the phenomena with the hope to be able to derive appropriated methods for remediation. Optic and/or auditory deficits of the eyes or the ears are easy to diagnose, but they have failed to give a satisfactory solution. It is in the brain, where one has to search for the deficits and – hopefully – for solutions.

From a neuroscientist's point of view the problem of learning at school may start in the sensory systems (not in the sense organs): we do not see with the eyes and we do not hear with the ears, but with the brain. When it comes to reading and spelling, the question is: what are the brain functions that are needed for fluent reading, what is needed for correct spelling?

During the last 30 years progress has been made in many respects concerning visual and auditory information processing and in the control of eye movements. It became clear that the processing is rather complex and that many different structures of the brain are involved reaching from low level subcortical centers to cortical areas.

The methods, that have been used to study the visual, auditory and optomotor system are mostly installed in laboratories using not only computers but also other equipment, which is not available for every day use. The data collected in these laboratories can hardly be reproduced in another laboratory unless one uses exactly identical equipment. Therefore, standard methods that can be used to investigate the age development of certain aspects of perception and/or of the control of eye movements have not been available until recently.

Even if one would find the standard methods for one aspect, say for the examination of visual skills, one would still want to have also standard methods for examination of auditory skills, such that a single child can be examined for (ideally) all possible deficits that might contribute to the learning problem.

About 15 years ago the optomotor laboratory at the University of Freiburg also started research in the field of eye movement control and its possible role in reading. It was found, that a specific component of controlling the fast stepwise jumps of the eyes (saccades) was systematically affected in children with dyslexia. Not all dyslexics suffered from this kind of optomotor deficit and the question came up, what the problem was with the rest of the dyslexics. The answer was found by using standard methods of the examination of auditory functions. The result showed, that deficits in the auditory domain are encountered even more frequently than deficits in the optomotor domain and that there are dyslexics suffering from both kinds of deficit.

Corresponding questions were asked in the case of problems with arithmetic. A fundamental visual capacity – subitizing and number counting by memory – was examined systematically by corresponding standard methods in normal control children and in children with dyslexia and with dyscalculia. Systematic deficits were found in both groups. While it was already an advantage to have the new diagnostic methods, the ultimate challenge remained open: what can be done, when deficits were identified. Is it possible to help the child to overcome the deficit with the hope, that learning of reading, spelling and/or arithmetic would be facilitated?

The optomotor laboratory looked for solutions for this therapeutic problem. Training by daily practice of visual, auditory and/or optomotor tasks were developed and evaluated. The results were encouraging: quite large percentages of the children were able to improve their skills within some weeks of daily practice.

The final question then was, whether or not the newly learned skills would help in reading, spelling, and/or arithmetic.

This book describes the methods and the results of the investigations of large groups of subjects in the age range of 7 to 17 years recruited over many years of research. The normal development, the deficits, the training, and the transfer of the training are described in detail and in a quantitative way. What is usually collected from many different research groups each specialized to a narrow band of the whole problem, has been accomplished by one group of scientists and their co-workers over many years.

In the field of dyslexia opposite points of view are often encountered. Many of the contradictions disappear when one looks at the data in a quantitative way and if one takes into account the different developmental stages of the children under investigation.

Since we are dealing with relatively low levels of brain functions, all the methods described in this book do not require any language processing. This implies, that they can be used everywhere in the world and hundreds of thousands children struggling at school may profit from the diagnosis and the training. The aim is not healing of dyslexia or dyscalculia. The aim is to "repair" identified deficits, at least in part, and to make it possible for the child to receive an education at school at an adequately intellectual level.

Therefore this book does not necessarily add to the neurobiological nature of dyslexia or dyscalculia. It is not written for scientists. It is written for readers who are concerned about learning problems of unknown origin: physicians, teachers, therapists, and parents.

The tremendous amount of data collected over the years and their analysis needed the cooperation of many co-workers.

On top of the list is Dipl. Phys. Klaus Hartnegg, who accompanied the work of the laboratory over many years. He did all the programming of the test and training instruments, he provided special programs for the collection and the analysis of the data. His work is the most valuable contribution to this book. He also critically read the manuscript to minimize errors and prevent misunderstandings.

Of course, most of the data have been collected for academic purposes like Diploma and/or doctoral theses (dissertation) in Medicine, Psychology or Biology. I am grateful for the contributions of Monica Biscaldi, Benjamin Fischer, Christine Gebhardt, Stefan Gezeck, Andrea Köngeter, Annette Mokler, Tina Schäffler, Juliane Sonntag.

The help of Peter Johnstone in contributing to the introduction and to the general aspects of development is greatly acknowledged.

Burkhart Fischer Freiburg, May 2006

General Introduction

Neuroscience is one field of the natural sciences that has produced millions of items of experimental data. It has provided a lot of knowledge about the basic physical and chemical processes in and between nerve cells. However, neuroscience has provided little understanding of the functional principles that make our brain what it is: an organ with a huge memory, which finds relationships between the contents of memory, which is able to learn functions, even for those for which it was not constructed, e.g. reading. (By function we generally mean a process, which relates specified input to specified output. In the case of reading the input consists of a series of visual signals (grouped letters) and the output is a spoken word.)

There is hardly another field of natural sciences, in which the discrepancy between the amount of data and the understanding of what one wants to explain, is greater than in neurosciences. In fact, the understanding of our brain is one of the last great challenges of sciences. For example, we maintain, that human beings are different from any other living species (animals), but it becomes more and more difficult to tell exactly, what the difference is.

In almost everything we do in everyday life we use brain functions that were not available at birth. We learned these functions in our early years without instruction. The brain has its own methods of learning: "trial and error" and "repetition". At first glance these two principles seem sufficient for learning, but there is another important factor. The goal of the learning process must be clear from the beginning or must become clear during the learning process. Problem solving is dictated in many cases by the biological goal of survival. When an organism struggles for survival it repeats its actions many times in search of a solution. Those who come to a solution will survive in the long run.

In affluent societies, survival does not play such an important role unless the individual or society is under threat. But the brain still applies its approaches to learning – trial and error and repetition – even in cases where biological survival in not the goal, such as in reading or spelling.

Repetition is essential in the acquisition of movement and skills, for example, in sport or music. However, repetition alone is not always sufficient for improvement to take place. If a musician is to play a passage correctly then he or she must have a correct interpretation in mind, perhaps supported by a teacher. What follows is a dynamic series of repetitions that increasingly approximate to the desired goal. Repetition involves an intelligent, dynamic sequence of trial and error, in which the successful trials are retained and the failures discarded. Repetition, in order to bring about learning, involves a level of insight or cognition.

As we will see, there are cases where the level of insight can be relatively low by knowing nothing but "right" or "wrong".

Repetition and rote learning were established methods of learning for hundreds or even thousands of years. Today it is sometimes said that these methods have become unfashionable in the search for better ways of learning. However, it can equally be argued that repetition and rote learning remain important in education. Now they are accompanied by insight and trial and error. As a result a pupil can apply the knowledge that he or she has acquired through rote learning not just to parrot it as meaningless information. Training and repetition are important tools in education.

This book deals with the diagnosis of deficits and the effects training (i.e. repetition of specified tasks) in the domain of auditory and visual perception as well as in the control of eye movements. (We will define later, what exactly we mean by "deficit"). It is important to understand the relation between the training of movements and the training of perception. From inside the brain both the control of movement and perception rely on nerve cell function. The functions of nerve cells and the functions of group of nerve cells is a matter of contact between the nerve cells. When training the control of a specific movement we activate these nerve cells with the intention of improving their contacts and of suppressing nerve cell activity that disturbs the execution of the desired movement. The same thing happens in our brain in the training of a specific perceptual task as in the case of movement, namely, certain groups of cells are activated or suppressed. Over time these assemblies of nerve cells are able to cooperate more effectively and to facilitate the perceptual process. Therefore, the basic principles of learning can be applied to movement and perception. Eye movements need separate consideration because they involve movements that are also essential for visual perception.

The significance of initiation and execution of eye movements in vision has been neglected for decades because these processes rarely reach our consciousness. Since the discovery of the optomotor reflex, known as express saccade [Fischer and Boch, 1983]; [Fischer and Ramsperger, 1984], the understanding of the close relationship between the visual and the presaccadic processes has increased. The saccadic eye movements and especially their reaction times were investigated by many groups in the field of eye movement research [Fischer, 1987]; [Fischer and Weber, 1993]; [Goldberg and Colby, 1992]. The complexity of the saccade control by the various cortical and subcortical structures was understood more fully and eventually provided ways of assessment, that can be used for diagnostic purposes [Munoz and Everling, 2004].

We are unaware of the processes of auditory and visual perception, including the use of saccades. We do not feel the brain activity accompanying perception. It seems to occur as a passive process that happens without our conscious decision and without any effort. As the subject being examined cannot report back on these processes, a diagnostic gap has developed between the sense organs (the eyes and ears) and the result of perceptual processes. This diagnostic gap could be entitled "preconscious perceptual processing" or "pre-perceptive processing" and analogy to "pre-motor processing".

Most of this book contributes to filling this gap. In addition, not only are the diagnostic aspects covered but also the therapeutic possibilities are described insofar as they are already established and evaluated. It will also be shown that successful training of functional deficits transfers to learning at school.

The Fig. 1 illustrates schematically the functional subdivision of the sense organs (the ears and the eyes), the sensory processing, and the cognitive functions including language processing, reading, spelling, and number calculation.

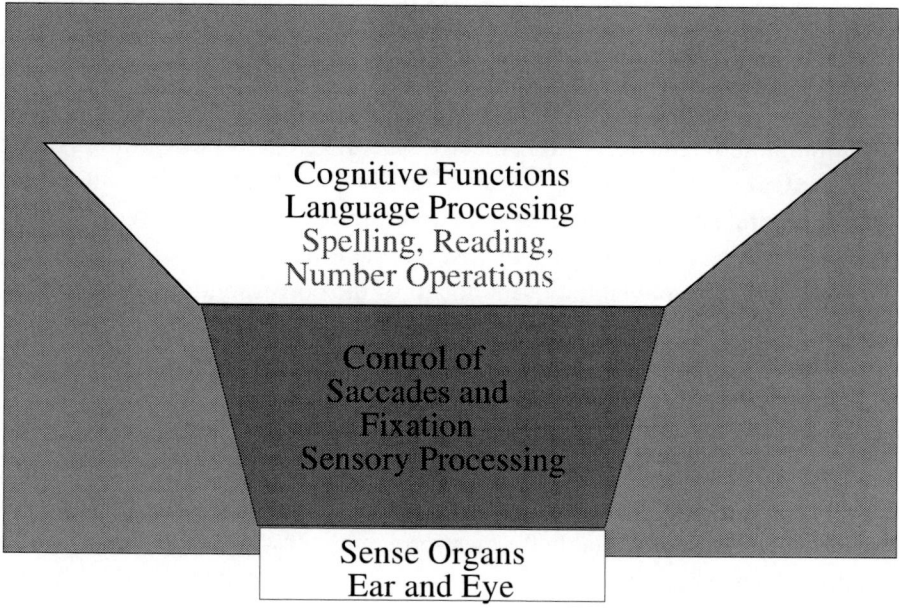

Figure 1. Schematic diagram to illustrate the functional subdivision of the sense organs, the sensory processing, and the cognitive functions.

The book deals with auditory and visual processing as well as with the control of saccadic eye movements and fixation, both being part of the visual process used in every day life and also especially in reading.

The part Development covers the sensory processing in normal subjects between the age of 7 and 55 or even 75 years. It describes the methods and the variables being used for diagnostic purposes. Each variable will be defined and it will be explained how the variable is measured quantitatively.

The part Deficits selects groups of subjects in the age range of 7 to 17 years suffering from specified learning problems: dyslexia, dyscalculia, attention deficits. An extra group is characterized by general learning problems: the members of this group could not be classified into one of the other groups. Most of them had low intelligence and/or other developmental deficits excluding them from the other groups by the means of diagnostic criterions. For each group we go through the auditory, visual, and optomotor aspects separately. The reader will be able to quickly look up the deficits occurring in one or the other group.

The part Training is organized according to the methods of auditory, visual, and optomotor training procedures, which are the same for the groups treated before separately . This makes it possible to find the success rate of each of the training procedures.

The part Transfer treats the 3 training procedures and their transfer to selected skills learned at school:

- The transfer of the auditory training to spelling

- The transfer of the visual training to basic arithmetic

- The transfer of the saccade training to reading

To make it easier for readers, who interested only in some specific aspects treated in this book, the table of contents gives detailed information to find the parts about Development, Deficits, Training, and Transfer with respect to the different deficits in the Auditory, the Visual, and the Optomotor System. Because one or the other reader wants to read only selected parts, several statements and explanations are repeated at different.

Almost all facts presented in this book rest on scientific studies and/or basic knowledge of neurobiology and neurophysiology. But not all of the corresponding references could be included. The most important ones are given in the text and the list of references allows the reader to get more detailed information. As far as review articles are available they will also be cited in the text.

The special advantage of this book may be seen in the fact, that a number of different functions of perception are tested in each member of large groups. The effects of training in all these different domains are treated using standardized methods. The diagnostic and the therapeutic methods are available for everyone, who may need them, because the steps, which lead from basic neuroscience to application, have been taken successfully during the last 15 years.

In the literature the results of scientific studies are always presented quantitatively by numbers, sometimes figures are added. In this book almost all results will be presented by figures: the reader can see at once, what the answer to a given question is. Numbers need translations in our brains, before we understand, what they are telling us. Figures use the visual system. They need some explanation also, but what they tell us becomes clear immediately. Therefore, this book can be regarded as a Figure Book more than a Text Book.

PART I

DEVELOPMENT

Summary

After a general introduction on human brain development the chapters of this part describe the development of auditory, visual, and optomotor functions from the age of 7 years to the adult age of 18, and further on with age increasing up to 65 or at least to 50 years. The methods for examination of these brain functions and their neurobiological background will be explained. The various variables provided by each of the tasks will be presented as age curves of the mean values and their standard error. The data will be used as the basis for the diagnosis of children with problems in learning at school. Therefore the first part of the book will be relatively long as compared to the other parts. As a general result we will see, that the age of development is not completed at the beginning of school, but lasts until adulthood.

Chapter 1

An Overview of Human Development

Summary

This chapter briefly looks at some of the factors in human development – maturation, learning, the phylogenetic and ontogenetic influences on brain development. It also outlines why human beings are more predisposed to communicate through spoken language than written language. As a consequence we find learning to read and to write difficult at school. We need some years of competent instructions and practice to learn this biologically unusual capability. To speak a language is learned much easier and much earlier in life.

1.1.   Introduction

The human species has developed over millions of years. We believe that human beings are the most advanced of any living species because our brains are capable of highly developed cognitive functions not available to other animals. The most prominent of these is language. While animals have also many ways of communicating amongst themselves, they do not have the capacity to learn the complex processes of producing and understanding the acoustic signals that constitute language. One may argue that thinking is internalised speech and develops only after the acquisition of spoken language. Young children exist mainly in the present, the here and now. The development of speech and language throughout childhood lays the foundations for abstraction, inference, deduction and thought that characterise adult behaviour.

When we talk about human development we should never forget that we are part of the biological world, and, that we are subject to the same biological laws as other species, animals and plants. From birth, babies can only survive with adult help. It takes 10 to 20 years for infants to mature and reach adult levels of functioning. This long process is evident when we consider, for example, the growth, development and maturation of our bodies from birth or even from conception to adulthood.

1.2.   Brain Development

It is tempting to believe that the development of the brain is completed in childhood. The number of nerve cells and their density does not seem to change very much after the age of 6 years. Even when examining the brain through a microscope, it is difficult to identify differences between the cortical structures of 10 and 20 year old subjects.

This is a premature conclusion because the development of the brain is not just programmed by our genes; it is also stimulated by learning. Brain development is the result of both genetic unfolding and learning. While the development of brain functions depends on the number or density of nerves cells, it is the density of the synapses, the contacts between nerve cells, which is significant. Brain functions are the result of cooperation of thousands of nerve cells via their connections. The density and functioning of the synapses is determined by learning processes. Learning means to improve the reliability of functioning of the synapses.

When we talk about the cognitive functions of reading or writing we accept that these are learned. But when we discuss hearing or seeing we may forget that these have cognitive components which have to be learned too. Many of the dynamic and cognitive aspects of seeing and hearing are acquired before schooling and without formal learning. The infant's world is "a booming, buzzing confusion". Preverbal infants and young children learn to discriminate, focus and respond by manipulating their environment and by experiencing the world through their senses.

Children learn to make sense of the world, but some may not process information with sufficient accuracy. The effect of this malfunctioning may not surface until later when higher order cognitive functions are involved in activities such as reading, writing and arithmetic. If the foundations for learning, the auditory, visual and optomotor functions are false, then some of the higher order cognitive functions may go off kilter. Consequently, even the existence of a high density of synapses does not guarantee the desired quality of brain functions, the synapses must also function correctly.

1.3.   Maturation

Genetic programming or maturation controls the development of our bodies and brains. This process can be compared to building a computer. Just as the components of a computer have to be connected in a certain sequence to each other, and, just as each major component, like the hard disk, has its own internal micro systems, so the body consists of interconnected major systems, such as the respiratory, nervous and digestive, each of which is has its own subsystems. Both the macro and micro systems of the body undergo an interdependent process of maturation. Maturation requires adequate biological conditions such as oxygen, temperature and energy supplied by food. In the nervous system the process of generating specific types of cells from single cells must be completed during pregnancy and the first years of life. The differentiation of nerve cells for generating and controlling movements in the motor system must also take place. Similarly, the sensory system must be equipped by specific nerve cells. The brain needs the adequate stimulation to establish the appropriate connections.

Maturation is a complex sequential, inter-related and hierarchical genetic unfolding of the different biological functions of the body. Each specific system becomes a building block for the sequential development of others. Specific processes (such as cell development in the nervous system) may take 2 to 4 years to mature, while the maturation of the human being, the summation of individual maturation processes, takes 25 to 30 years.

1.4. Phylogenetic and Ontogenetic Development

Ontogenetic development deals with the processes of development within an individual's lifetime. Phylogenetic development deals with the genetic and biological adaptations of a species over hundreds of thousands, or even millions, of years.

Mutations in the genetic material of a species result in an extremely slow process of selection of responses and adaptations that help the species survive. Our brains do not have a reading or spelling centre because these activities only recently made an appearance in human history, not long enough to bring about genetic changes.

It is not therefore surprising to find that we share some basic brain functions, for example, our visual and optomotor system, with other animals such as human and non-human primates. The developmental period for of the visual and optomotor system is about the same of other physiological functions such as the uptake of oxygen by muscle cells. Once these functions have matured they work at optimal levels until deterioration takes place relatively early in life from about 35 to 40 years of age.

A biological human generation covers a time period of about 30 to 40 years. The fact that our lives last much longer, up to a factor of 2 or 3, is the result of medical and technical progress made during the last 200 years. But these progresses are hardly reflected in our physiological functions: there was no time to profit from any evolutionary selection based on chances in the human genetic material.

1.5. Language and Written Language

Language developed relatively early during the phylogenesis of the human species. A detailed communication system between members of groups helped survival. Written language, on the other hand, was developed only recently, about 5000 years ago. People drew small pictures to represent what they saw. Their brains were able to see, interpret and remember those pictures. Our brains are prepared to see objects and they are able to recognize a small picture of objects. Written language using letters was developed only "yesterday" on a phylogenetic scale.

Letters are artificially selected visual signs used to record and recall that which has been expressed in voice, spoken word and thought. It is not easy for our brains to learn such an artificial construct as written language. Knowing how to use letters has never been a criterion for survival. Even today we have millions of illiterates. They do not need a written language for survival, they need water, food, and medicine. Even in the highly developed industrial states there are thousands of illiterates.

If one looks at the ontogenetic development of language and written language, one immediately recognises the differences between the two. Children begin to speak and understand words from around the age of 2 to 3 years. They do not need school lessons; they learn it almost by themselves, even most grammar is used correctly. By 4 or 5 years of age children have a good grasp of spoken language. By contrast, written language requires instruction. There is no natural development of written language and it requires the support of structured learning. Children need years of schooling to read and write.

Some children find the acquisition of reading and writing difficult. They have learning problems specifically related not to the spoken, but to the written language. From a neurobiological point of view one may be even surprised that most children are able to learn such a complex and intricate task as written language. It is not surprising that learning to read and write causes some children problems.

1.6.   Dealing with Numbers

Numbers have been important for the development of mankind, for example, the number of enemies, the number of family members, the number of deer are important calculations for survival. Simple operations, like adding the members of two families, may have been important but not everybody had to be able to do these operations. We know that ancient cultures used complicated mathematical systems. The use of money to buy products, instead of exchanging products by bartering, prompted the development of simple number operations within cultures. Like speech, the concept of number may have developed relatively early during the phylogenesis.

Consequently, we are not too surprised to discover that young children, and even some animals, have a concept of number. They may have a basic visual function to know how many items they see and to be able to tell the difference between (small) numbers of items.

Numbers play an important role in contemporary life. We count money and we "count" the time in numbers throughout the world. Yet, having a concept of number is more than just dealing with numbers. Without a concept of number even simple operations become difficult or impossible. We will see that a specific visual function can be used to develop a concept of number.

The interested reader is referred to the excellent book of Dehaene, who describes all these different aspects of the sense of number [Dehaene, 1997]; [Dehaene, 1999]

1.7.   Working Memory

Basic brain functions are coordinated with each other and rely on a memory that contains, for example, the information of a sequence of movements. This memory cannot contain all the information of all the possible actions we can initiate. It is sufficient for all this information to be held within a relatively short-time window, say a few seconds up to a minute. For example: after you have filled your cup with coffee, you are not supposed to turn it upside down, when you walk to your desk. This memory is called the "working-memory". The functions of the "working memory" are located in the frontal brain. Animals also need a working memory for their actions. There is only a quantitative difference between animals and humans with respect to their frontal functions. This part of the human brain developed relatively late on the phylogenetic scale and, on the ontogenetic scale, it is programmed relatively late: it may start at the age of 1 or 2 years and lasts up to the adult age. The frontal brain functions develop by learning. A component of saccade control is among these frontal functions. Neuropsychologists have developed a battery of tests to examine the functioning of the working memory in human subjects and, in particular, in patients with brain lesions.

In the analysis of the different sub-functions of perception in the following chapters we should remember that working memory is implicit in the visual and auditory functions being examined. When considering the generation of fast eye movements (saccades) we will see, that a specific component relies on an intact frontal lobe.

The reader interested in the frontal brain is referred to the book "The Prefrontal Cortex" [Fuster, 1991].

1.8.   Development of Motor and Sensory Systems

Young animals and human babies have difficulty controlling their movements. This is partly due to immature muscles, but more serious problems arise from the control of motor functions by the brain. The nerve fibres in the brain are still not completely myelinated until the age of 2 or even 3 years and therefore the delivery of messages to the muscles is still slow. Also, motor control still suffers from an insufficient input from the motor cortex. We are able to easily observe this motor control improving over the years by maturation. For example drawing begins already at 3 or 4 years but the eye hand coordination is still far from adults. Even at the beginning of school age (6 years) this sensory-motor skill needs still more practice until it reaches an adult level. An other example is handwriting, which keeps changing over many years until a typical personal form is reached which may remain the same for the rest of life.

While motor actions can be observed, which makes it easy to see the developmental improvements, it is harder to analyse the development of optomotor control and sensory systems. We may infer from our observations of the development of movement that the sensory system develops concurrently in a similar time scale. On the other hand, we may believe that the development of the sensory systems reaches its final level earlier. For example, visual acuity reaches values close to 1.00 at around 7 years of age. Auditory detection thresholds reach their adult level already at the beginning of school age, i.e. at 5 to 6 years of age.

However, the more complex tasks that test the brain functions that serve the performance of perception may show a different sequence of maturation. The tasks considered in this book challenge brain functions, not the functions of the sensory organs themselves. These functions concern the necessary processing of the sensory signals on their way to conscious perception. Most of these functions remain unconscious. This is one of the reasons why they have been neglected for such a long time in the diagnosis of visual and auditory capacities. Consequently, until now, there were no diagnostic instruments and standard tasks available for the specific examination of signal processing and their possible role in learning problems.

Chapter 2

Development of Auditory Functions

Summary

This chapter describes the development of auditory functions, which are necessary for almost perfect decoding of the auditory signals that contribute to the understanding of words. The issue is not, whether or not subjects are able to hear a sound at all, but to be able to distinguish between similar sounds. The corresponding functions must not include any language processing nor should they require more than normal intellectual capacities. We will consider 5 independent sub-functions, which are considered as basic, when analysing words. The methods, that form the basis for the diagnostic tasks, will be explained in this chapter. The age development will be shown by the age curves of the corresponding variables.

2.1. Introduction

The auditory system begins to work in full from the beginning, even before birth. However, at this time the functions were quantitatively still in an earlier mode of development. High quality auditory processing is needed by the time a child begins to speak. The learning process is one of the type replicating by the own voice what was the result of the auditory process. This is kind of a circling process which needs to be repeated until a satisfactory result is obtained: what the child produces by speaking sounds about the same as what the child heard. The development of language depends on the development of the auditory system and vice versa.

When it comes to the learning of a written language, the demands of the auditory system increase. It is no longer enough to have a certain amount of congruence between the voice (words) one hears and the voice (words) one produces. One has to hear the details of the pronunciation of the word in order to know the spelling. Of course, not all the details of spelling can be found by correct auditory analysis, but it is of great help, when all the auditory support is available.

These facts have been know for long time and tests have been used to examine the capacity of auditory discrimination of similar words in a given language. However, such tests also challenge the quality of the knowledge of the language, from which the words are selected. Even if one reduces the test words to syllables these are still parts of the language used. One tries to examine the subjects ability to work with auditory "material" irrespective of the context of meaningful words.

During recent years this auditory capacity called "phonological awareness" has been become popular as an important aspect of the relationship between spelling and audition. Unfortunately, there is no unique standard method to examine this capacity and therefore a sufficiently clear definition is missing. Different methods have been offered and, no wonder, they produce different results.

The afferent auditory system forms a complex network of nerve cells, their fibres and structures. The Fig. 2.1 shows a schematic drawing presenting the most important anatomical structures and their connections to the auditory system. The systems for the left and right ear are both shown separated by the vertical broken line. One sees quite a number of lines crossing the midline thereby connecting the systems originally driven by the sensory input from one ear to the processing of the other ear. The red lines represent information travelling from the periphery towards central brain structures, the green lines indicate information travelling from the central structures to the periphery modulating the peripheral processing by results of central processing. Even though only the lines from the left ear are shown, the diagram is already rather complex. With the lines from the opposite side we would not see anything anymore.

2.2. Low Level Auditory Functions

Here, we consider, what is called "low level auditory discrimination". "Low", because no language processing is included at all. It was long known that the auditory signals are processed in a complex network of nerve cells after they leave the receptors and primary nerve cells along axons that form the acoustic nerve and before they arrive at the auditory cortex.

The Fig. 2.1 shows that the processing does not take place in the ear as a sense organ and it does not take place at the cortical level of language processing. Is it possible, that problems at this intermediate low level contribute to problems in spelling or any other cognitive capacity to be learned at school?

2.3. The Auditory Tasks

The diagnostic tasks come down to looking at some basic physical acoustic properties of spoken words.

Two of these properties are straight forward: intensity (or volume) and frequency (tone pitch). Another is a very short temporal gap between two different sounds. In spoken words such gaps occur whenever the stream of the acoustic signal is briefly interrupted. The detection of each single gap helps to analyse and identify the word. The temporal order of a sequence of physically different sounds should also be recognized correctly. This aspect concerns the speed of auditory perception. If sounds follow each other too fast, one can no longer keep them apart. If a subject's auditory system is too slow, it cannot keep the sounds in their temporal order. Fusion or confusion is the result. Finally, one also needs

Auditory Processing

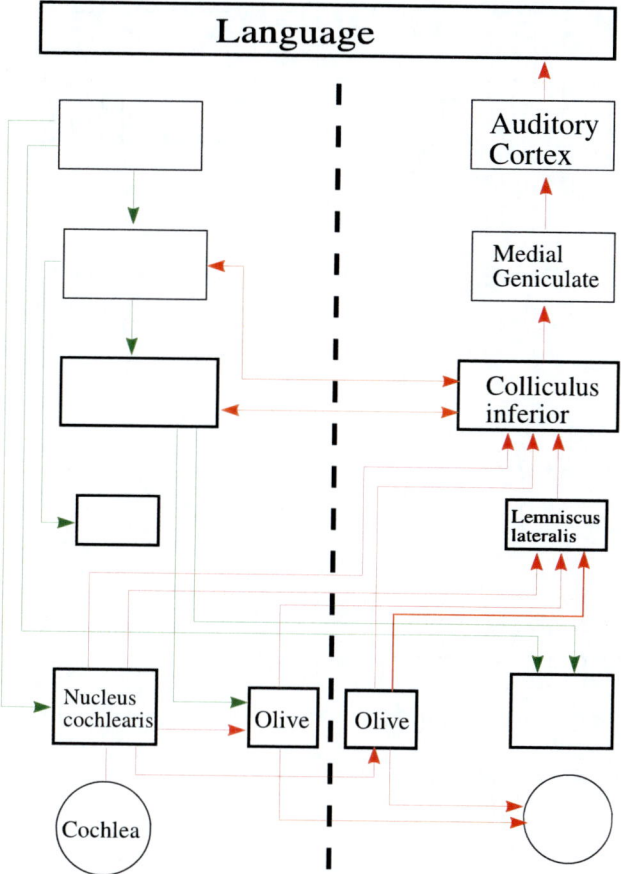

Figure 2.1. Schematic drawing of the auditory system. The left and the right systems are shown both, because they have several interconnections involved also in centrifugal fibres. The anatomical structures are shown by the boxes with the names inside. The broken line indicates the midline. Red lines are sending information from the periphery to the central structures, green lines carry information from the central structures to the periphery.

to keep apart signals that reach the two ears at different times. We call this function: side order.

The Fig. 2.2 shows schematically the design of the 5 tasks.

All 5 discrimination tasks were based on a two alternative forced choice procedure. Two stimuli were presented one after the other. The subjects were required to press one of two keys (two-alternative) corresponding to their perception. Subjects were asked whether the second stimulus was louder (was higher, contained the gap, was higher) than the first stimulus. If this was the case, subjects had to press the left key. Otherwise they had to press the right key. Subjects were required to wait until the second stimulus was presented before pressing the key. No feedback of correct or incorrect key presses was given during the test

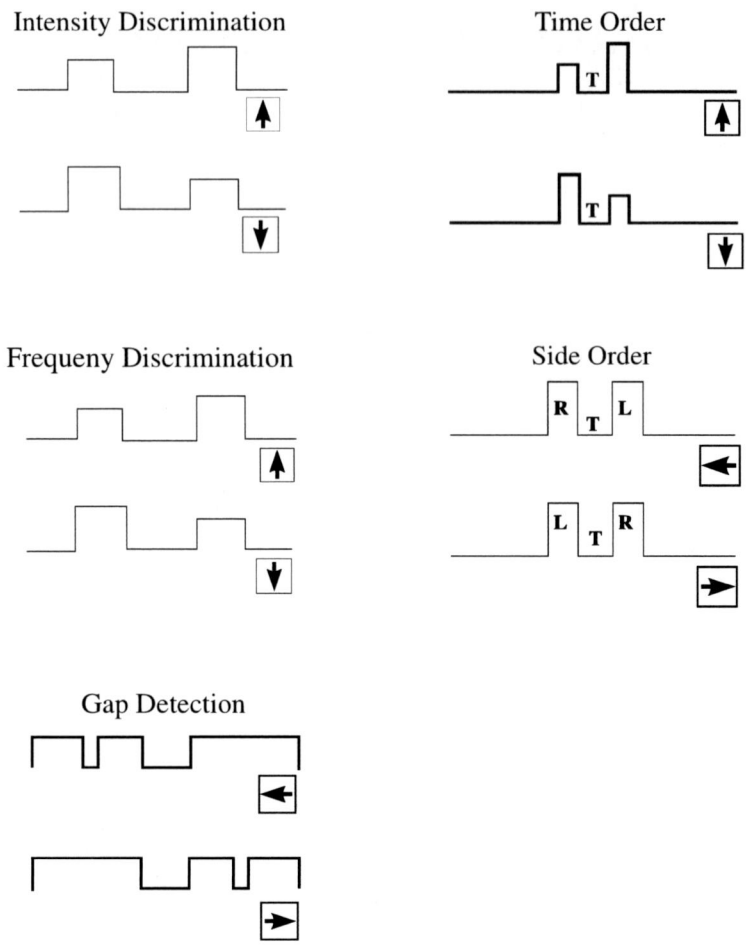

Figure 2.2. The panels illustrate the design of the 5 tasks to examine low level auditory discrimination. Each task is a two-alternative-forced-choice task. For each task both possibilities of stimulus sequence are shown. The detail of the tasks are explained in the text.

sessions. For the side order task the left had to be pressed when the second stimulus was presented to the left ear and vice versa.

The difficulty of each the task was increased within each session by decreasing the difference between the two stimuli in steps of decreasing size (see below). For each task the threshold value was defined as the last correct response preceding the first of 3 errors in a sequence of 7 consecutive trials. This method was used to find the transition from series of "almost all correct" responses (98%) to a series of "50% correct" responses. The algorithm does not imply that only 7 trials were used to determine the threshold, because errors preceding this sequence may have occurred, but did not terminate the session. Each of the 5 sessions lasted about 4 minutes depending on the child's performance.

It was not the aim of this method to determine the threshold exactly, but rather to have a fast routine test for single subjects and compare the individual results with those of an age-matched group of controls having been tested by exactly the same method.

Computer simulations of this algorithm showed that the turnover from 98% to 50% correct responses (the guess rate) is determined in 63,4% of the cases with a precision of ± 1 step. In 83,5% the error was ± 3 steps or less. On average the algorithm has a small tendency to judge the threshold "lower". This kind of precision can be considered as adequate given the much larger interindividual scatter in the data (see below). A more precise threshold determination would have required much longer testing times with the consequence of possible effects of fatigue and attention problems. The methods of the tasks are described in detail elsewhere [Fischer and Hartnegg, 2004]

The parameters of the 5 tasks are:

- Intensity discrimination: Two white noise tones 300 ms in duration were used. The interstimulus interval (ISI) was 150 ms. The intensity of the reference signal was 55 dB(A), each trial started with a test intensity of 63 dB(A). On each trial the difference between test and reference stimulus was decreased by 10% of its previous value.

- Frequency discrimination: The reference tone had a frequency of 1000 Hz, 300 ms long and had an intensity of 65 dB(A). The test tone started with 1100 Hz (same duration and intensity as the reference tone). ISI was 150 ms.

- Gap detection: The test tones had an intensity of 60 dB(A), they were 300 ms long and consisted of white noise. One of them contained the temporal gap. The two tones were identical in duration regardless of the gap. Gap duration started with 40 ms. ISI was 300 ms.

- Temporal order: A 1000 Hz tone and a 1120 Hz tone were used. They were presented in random order to one ear only. Subjects were asked to indicate whether the higher tone or the lower tone was presented second. Both tones were 200 ms long and had an intensity of 63 dB(A). The start value of the stimulus interval was 300 ms.

- Side order: Clicks with 55 dB(A) were delivered, one to the right one to the left ear in random order. Subjects were asked to indicate if they heard the second click from the left or right. Start value of the stimulus interval was 300 ms.

2.4. Procedure and Analysis

The test procedures were carried out using a custom made hand held device with an built-in response key pad. A small LCD screen provided feedback only during the instruction trials. Stimuli were applied through headphones.

Test-retest reliability was determined for all 5 tasks separately. The 5 test-retest correlation coefficients were calculated. All of them reached significance.The scatter plots showed that the deviations of r from the ideal value 1.0 were mostly caused by cases were the second value was "better" than the first. This indicates (and will be shown in the chapter on Training), that the repeated performance of these auditory tasks is subjected to learning effects, which transfer to spelling (see chapter on Transfer). It will also become clear from the results in the chapter Diagnosis that the reliability was good enough to differentiate between groups of dyslexics and controls.

The subjects were classified into age groups. The bin widths were increased by one year per bin: 7-8, 9-10, 11-13, 14-17 years, and so on.

2.5.   Age Curves of Low Level Auditory Functions

A number of subjects could not perform one or the other task better than by chance. All subjects from the beginning of the test trials pressed the key only by guessing. We called these tasks "unsolved task". They came as a surprise, because none of the normal adult control subjects (aged 20 to 30 years) had difficulties in any of the 5 tasks to reach reasonable threshold values. Therefore, as a first step of the data analysis we counted for each subject the number of these "unsolved" tasks.

Fig. 2.3 shows the age curve of this global variable of the auditory task performance. Note that only the adult group was able to solve all tasks. Among the younger and older subjects there were always some who could do all tasks, but the mean value of the group was not zero. Many children at the beginning of school were unable to solve 2 out of the 5 tasks. Only 20% of them could solve all 5 tasks. Interestingly, none of the subjects failed all 5 tasks.

One might argue, that the tasks were too difficult. But it is necessary to make the tasks difficult enough, that even the "best" test person has difficulties to perform them, to be able to differentiate their performance. (when the tasks were made easier by allowing really great differences between the two stimuli, the results were not systematically different.)

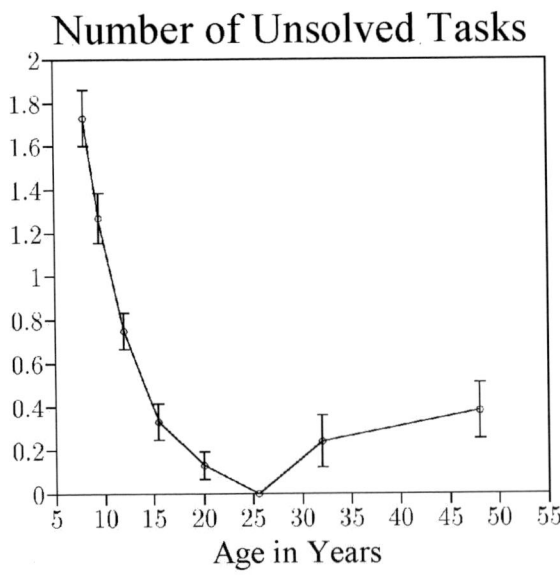

Figure 2.3. The figure shows the mean number of unsolved tasks as a function of age. The ideal value would be zero. Only in the adult group around 25 years of age all subjects could do all 5 tasks.

It is of considerable interest to see which of the 5 tasks were the most difficult ones. For each task the percentage of subjects, who were able to reach a reasonable threshold value was calculated from the data. The result is show in Fig. 2.4. It shows, that only the side order task was completed by more than 90% of the subjects. Among the teenagers and adult subjects all participants succeeded in the task performance better than by guessing. In the other 4 tasks the ideal percentage of 100 was reached not before the age of 18 years. The time order task was especially performed poorly by many subjects before the age of 20 and after the age of 30 years.

By excluding the unsolved task from the calculations, we can plot the mean values of the thresholds as a function of age in Fig. 2.5. A similar picture emerged for all 5 tasks: the threshold became lower with age from 7 to 20 years. Only in the side order task the thresholds remain as low as before when age increases above 25 or 30 years.

In principle it is possible that the 5 task challenge similar subfunctions or even a common auditory function and therefore knowing the result of one task would allow one to know the result of the others. But this is not the case. The 10 pairs of cross correlations between the variables of the 5 tasks were calculated for the largest age group (10 to 13 years, N=120). They reached only very small correlations coefficients r between 0.14 and 0.28. Some of them reached significance values p smaller than 1%, because of the large number of subjects. The only exception was the correlation between frequency discrimination and time order. The correlation coefficient was r=0.43 (p=0.001). While this result may not come as a surprise (both tasks deal with frequency) the data show, that despite this correlation it is not really possible to predict the result of one task on the basis of the result of the other.

Even in the case of unsolved tasks it was not possible to predict, whether or not a subject would be able to solve one task given the subject has solved or failed another.

Therefore, a poor auditory performance as revealed by the present tasks is not caused by a general deficit in attention or any other general deficit. One needs to do all 5 examinations to really know the result for a given subject. The notion of independence of the tasks will be further supported when we analyse the results of the training, by showing that one task may be learned by many subjects another by much fewer subjects.

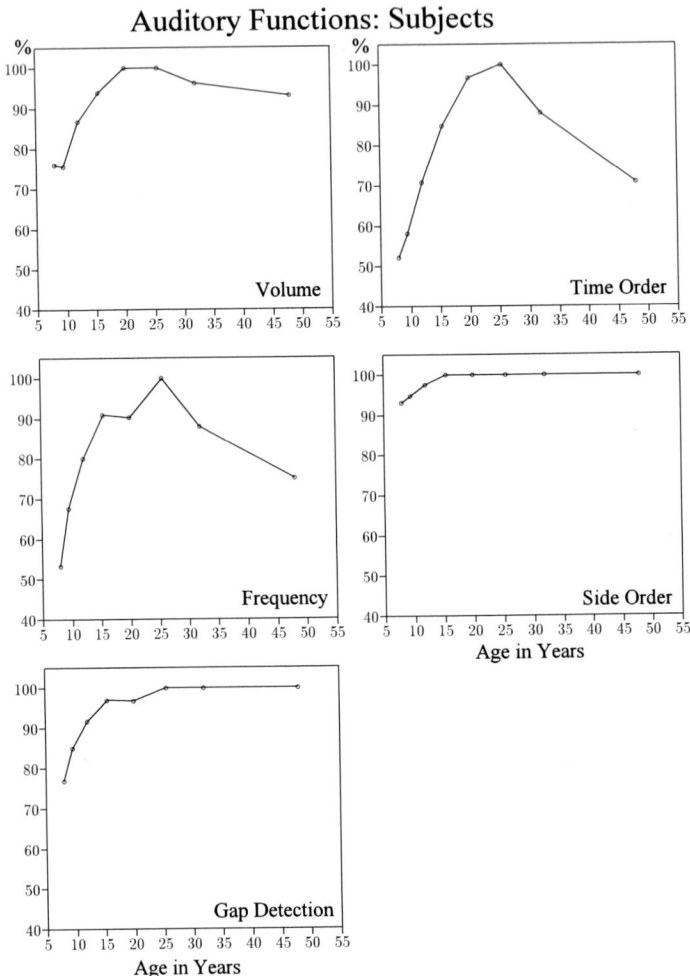

Figure 2.4. The panels show the percentage of subjects who could do the auditory tasks better than by guessing as a function of age.

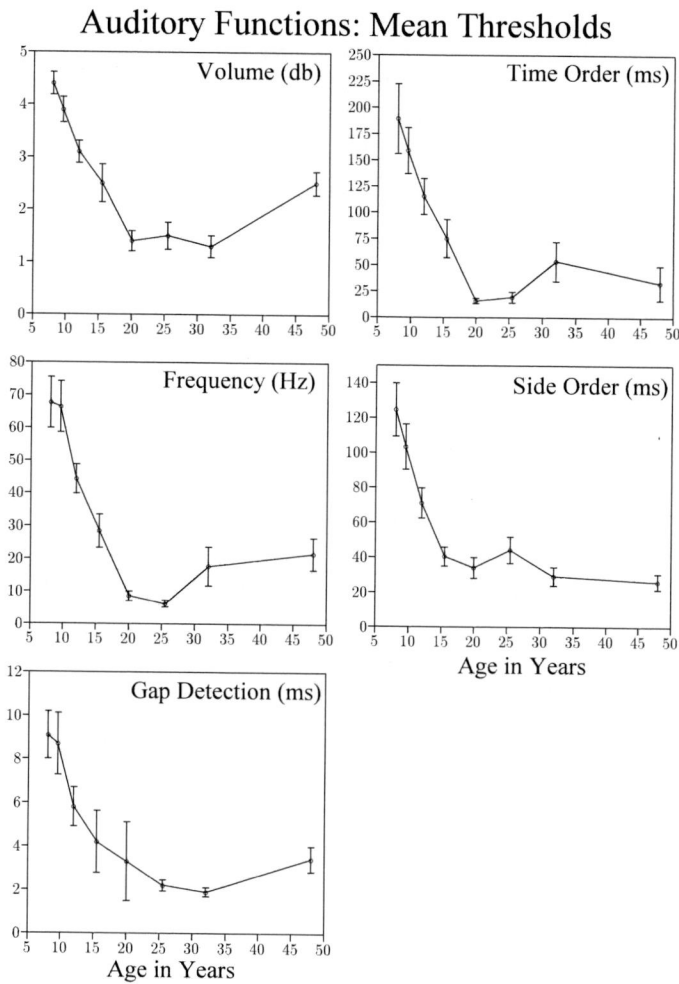

Figure 2.5. The panels show the age curves of the mean threshold values reached in the 5 auditory tasks.

Chapter 3

Development of Visual Functions

Summary

This chapter deals with the development of two special visual functions. Both have to do with temporal aspects of vision and one includes short term visual memory. Dynamic vision – as contrasted by stationary vision – concerns the speed of the processing of the signals in the subcortical and cortical visual centres. Subitizing and counting by memory enables the subject to recognize the number of objects when they were presented only for short periods of time. The methods used in this chapter will also form the basis of the diagnostic tasks used in the next part of the book. We will see, that the development lasts throughout the first and second decade of life.

3.1. Introduction

Vision is the most important sensory brain function for human. About half of the cortical surface of the brain is devoted to or involved in visual and optomotor functions. Involved cortical brain structures include the occipital, parietal, inferior-temporal, and frontal lobe. A number of sub-cortical structures serve in the visual processing before the cortical processing takes place, which eventually leads to a conscious perception. Other structures send visual information to the brain structures for eye movement control.

At birth the fibres in the optic nerve are still not myelinated and therefore the visual processes are still functioning in a very primitive way. The sensory systems have this in common with the motor system. But even when the nerve fibres are physiologically "equipped", the visual system is still far from having reached its final developmental state. The appropriate synapses are not in perfect action, the correct connections of the different parts of the right and left eye to the corresponding parts of the lateral geniculate nucleus and the visual cortex still must be established. These processes are part of the maturation, but they also need the support of the signals from the retinal nerve cells to the central structures. If these signals do not arrive at all or arrive in wrong order, the development of vision and of binocular vision is in great danger and it may happen, that it will not reach a functionally correct state at all. This is the reason, why the orbital position of squinting eyes should be corrected as early in life as possible.

The basic rules for the development of any brain structure implies, that babies and infants need to use their "eyes" as often as possible to see all kinds of natural visual stimuli. The brain learns and maintains only those functions that are needed and used.

The adult human visual system is a complex network of cooperating neural structures. The Fig. 3.1 schematically shows the visual system. Both, the lateral geniculate nucleus and the visual cortex are organized in 6 layers of cells. The layers 1 and 2 of the lateral geniculate nucleus contain the large cells (mango cellular layers), the layers 4 to 6 contain the small cells (parvo cellular layers). Even though both cell types project to layer 4 of the visual cortex, the two systems are kept anatomically and functionally separate and project to different parts of the higher visual cortical structures.

Each eye sends information into one hemisphere of the brain. The general rule that the left part of the body is represented by the right hemisphere (and vice versa) does not hold for the two eyes. Here the correct rule is: everything which is in the right field of view is represented by the left hemisphere and vice versa. This fact is very important, because otherwise the same object would be represented twice, once in the left and once in the right hemisphere. Note, that the cell layers of the lateral geniculate nucleus are strictly monocular and the m-system (m = magno) is separate from the p-system (p=parvo). It is only at the level of the primary visual cortex, where we find nerve cells, that can be activated by stimulation of either eye and that carry information of binocular disparity [Poggio and Fischer, 1977].

For the understanding of the optomotor control during reading it is most important to distinguish between the two subsystems of neural processing: the magnocellular and the parvocellular system. Both originate in the retinal cell layers. The magno-system consists of relatively large cells with thick axons and fast conduction velocities. The parvo-system consists of smaller cells with thin axons and relatively slow conductions velocities.

The two systems are well separated in the different layers of the lateral geniculate nucleus. Only the layers 1 and 2 receive input from the magno-system. At the level of primary visual cortex the two systems are also separate. From there the parvo-system projects to the infero-temporal cortex, where the visual system works for the identification of objects. This pathway is called the ventral pathway or the WHAT-system, because it tells us, what objects we are looking at.

The magno-system, on the other hand, projects to the parietal cortex and further on to the frontal lobe structures. This pathway is called the dorsal pathway or the WHERE-system, because it tells us, where things are in space [Mishkin et al. 1983]. The primary visual cortex also sends signals to the visual part of the tectum (the superior colliculus). The lower part of the superior colliculus sends signals to the brainstem centres, which finally innervate the eye muscles. For review see [Goldberg and Colby, 1992].

Therefore, when we talk about the optomotor system, we have to concentrate on the magno-system. Because of its physiological properties the magno-system has a relatively high temporal resolution, which enables the cells to "see" for example movement. But any other aspects of vision, which includes fast temporal changes of the stimuli, is processed by the magno-system. One can imagine already from these facts that seeing when the eyes move in many small steps 3 to 5 times in a second, one better has an intact magno-cellular system.

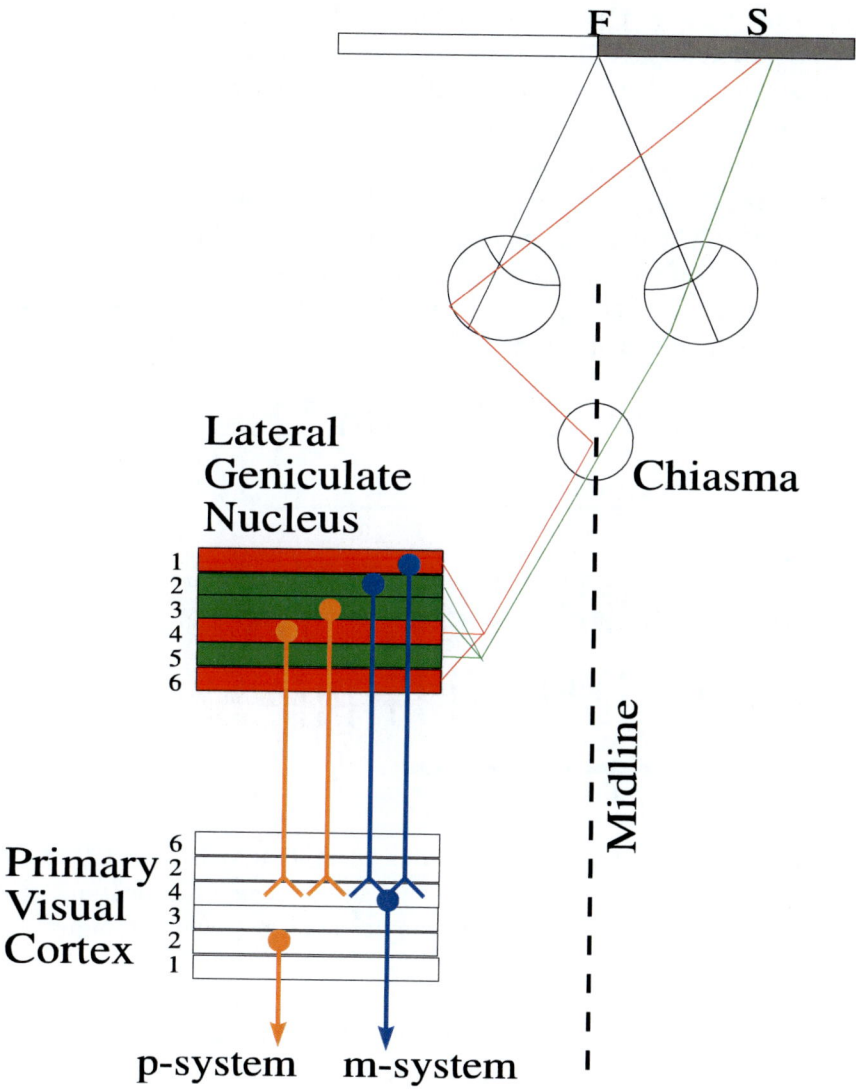

Figure 3.1. Schematic drawing of the main anatomical connections of the afferent visual system. Red and green mark the pathways from the left and the right eye, respectively, blue and yellow mark the projections of the magno and the parvo cellular system. Note, that the fibres of the left eye receiving information from the right visual hemifield do not cross to the opposite hemisphere at the chiasm. F indicates a fixation point, which is projected to the fovea in the middle of each eye. S represents a visual stimulus at the right side from the fixation point. The stimulus is "seen" by each eye. But because of the crossing and non-crossing fibres at the chiasm, the signals of both images are projected to only one hemisphere. For details see text.

The Fig. 3.2 shows the sequences of nerve impulses following the onset of a visual stimulus in the cells receptive field. The cells of the magno-system respond to the sudden onset of a visual stimulus by a short burst of nerve impulses lasting about 200-300 ms, called a transient response. Only a short time later they return to level of impulse activity that had before: the cells have "forgotten" that there was a stimulus event and are ready to respond to next stimulus. By contrast, the cells of the parvo-system respond to the sudden onset of a stimulus by an increase of their activity, which lasts as long as the stimulus is present which is therefore called a sustained response.

Yet, the activity of the parvo cells does not lead to a sustained perception. If we manage to stop all movements of the retinal image (for example by fixation our eyes in a dim room) we lose the perception of objects. In fact, the total field of view may become a grey surface. This phenomenon is called "fading" and it shows that changes, like movements of the retinal image are necessary to maintain vision.

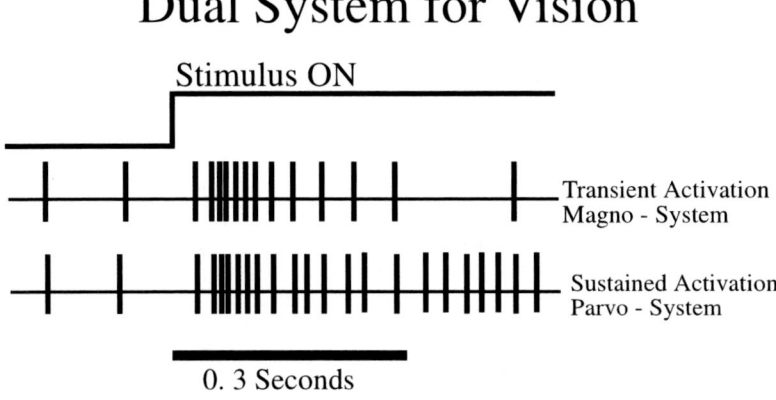

Figure 3.2. Nerve pulses as reaction to stimulus onset, time runs from left to right. Cells of the m-system respond to the onset of a visual stimulus by a transient increase of their impulse rate. Cells of the p-system keep firing impulses as long as the stimulus is presented.

Interestingly, in classic ophthalmology the factor of time does not play any significant role. The battery of visual diagnostic tasks does not contain a task which examines the speed of vision or the quality of movement perception. Yet, it has been quite clear for many decades that the visual processes in the brain takes time and that they are processes changing with time. We know that the visual system has a time limit, that allows it to present fast sequences of pictures which are no longer perceived as separate pictures. They merge into each other. If the pictures are sufficiently similar, but not identical, we perceive the sequence of separate frames as movement. This is the physical and physiological basis of cinema and of television. But if the sequences figures are too fast, vision collapses altogether, colours and contrast are fused.

The magno-system is responsible for sufficiently high speed of vision. It forms the neural basis of dynamic vision as opposed to static vision. The magno-system allows to separate the 3 to 5 pictures produced by the sequences of saccades sufficiently well. The separation, however, does not lead to a perception of displacement of the images: we still see a stable world.

Below we will describe dynamic vision, but not static vision, because we need this knowledge for understanding the visual processes in reading. In reading it is of high relevance that each retinal frame is correctly seen in the correct sequence and that no frame is skipped. Only highly trained readers can afford to skip words or parts of them, still knowing what they read.

The functional differences between the two systems and the significance for visual information processing and attention has been recognized relatively early [Breitmeyer and Ganz, 1976]; [Steinman et al. 1997].

3.2. Dynamic Vision

To examine dynamic vision one needs to use test stimuli, which change in time. There are several different possibilities, that may be implemented in a dynamic vision test. We have decided to use a test that is as simple as possible in several ways: simple enough to be understood by children, simple enough to deliver a quantitative measure of the quality of task performance, and simple enough to be built into a transportable test instrument.

Since the analysis of orientation is one of the most prominent features of nerve cells in the primary visual cortex, an orientation detection task was used. The Fig. 3.3shows the spatial and temporal aspects of the task.

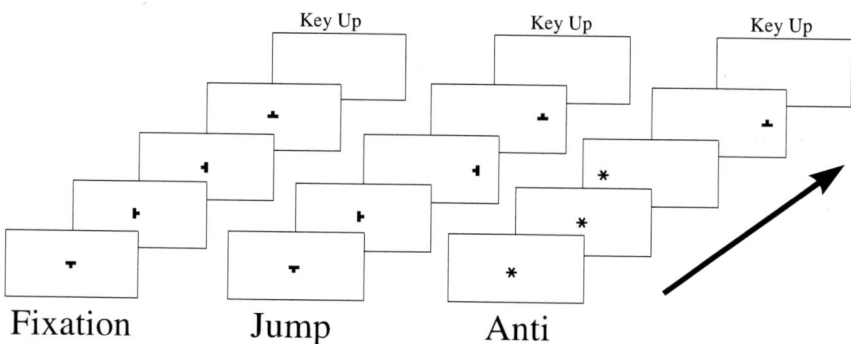

Figure 3.3. The figure shows the temporal sequence of frames in the 3 versions of the task to examine dynamic vision. Time runs from the front backwards. The timing of the frames is described in the text.

A symbol similar to the capital letter T is presented in one of the 4 main orientations (up, down, right, left) in fast sequences. The symbol was presented in black against a greenish background. The size of the test stimulus is 1.75 mm x 1.15 mm, corresponding to 0.33 deg x 0.22 deg at a viewing distance of 30 cm. The contrast is 0.8. With these conditions the stimulus is very easy to see and it is very easy to recognize its orientation as long as each stimulus can be seen for long durations (in the order of a second). Everybody with normal visual acuity recognizes the correct orientation with a hit rate of 100%.

The presentation time of each orientation was 180 ms, the frequency 5.6 Hz. A sequence consists of 3 to 5 presentations with randomly selected orientations and a randomly selected number of presentations, such that the subject does not know, when the sequence would be terminated. The task is to press the up, down, right, or left key of the test instrument corresponding to the last orientation the subjects has seen. This response could be right or wrong. There is no time constraint on pressing the response keys. The task was repeated in 100 trials. Since the best strategy for correct responses is to maintain fixation in the centre and to identify the orientation by foveal vision, this task is called the Fixation task. More details of the methods have been described earlier [Fischer and Hartnegg, 2002].

The result was presented as the effective percentage of correct responses, renormalized taking into account the fact, that the chance of hitting the correct key is 25%. Therefore, 25% correct responses resulted in a score of zero, 100% correct responses resulted in a score of 100%.

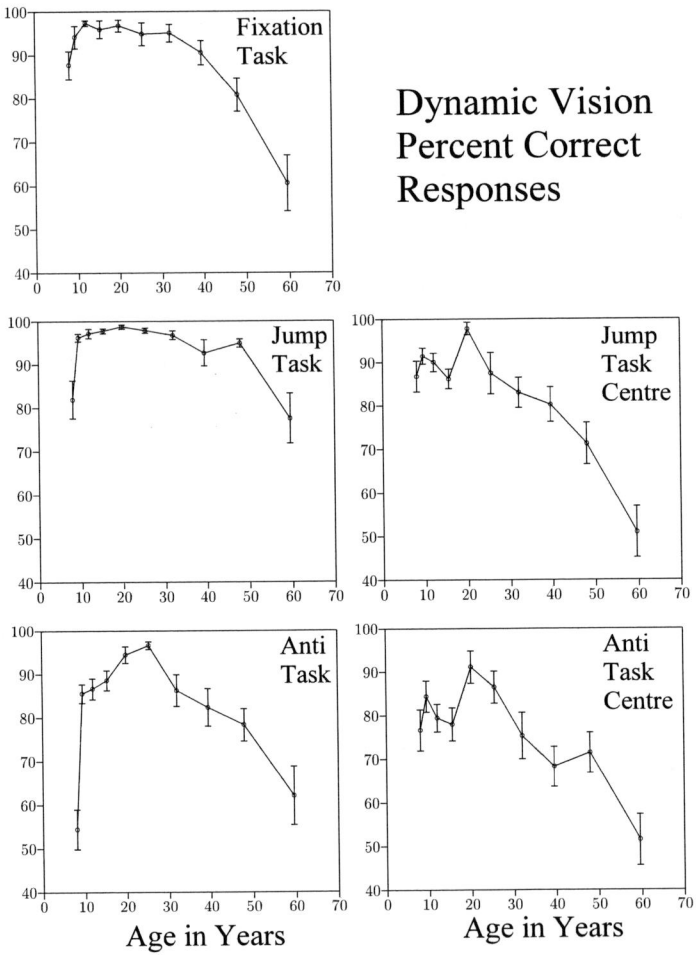

Figure 3.4. The panels show the age development of dynamic vision. Note that the best responses are given in the Jump Task, not in the Fixation Task. Note also the relatively low rate of correct responses in the Jump Task/Centre and in the Anti Task/Centre as compared to the Fixation Task. For details see text.

Fig. 3.4 shows the age curve in the upper left graph. Children at the age of 7 are performing relatively well with mean scores just below 90%. A further small increase up to almost 100% can be seen with increasing age. The performance stays at this almost perfect score until the age of 40 years, when it begins to decrease reaching values around 60% at the age of the mid-fifties. For some subjects it came as a surprise, that they believed they could not see the correct orientation. Yet, when they passed the test it turned out, that they performed quite well. The opposite also happened: the subjects judged their performance much better than it really was.

The task was then modified by suddenly displacing the rotating test stimulus from the centre to the left or to the right, in random order from trial to trial. The stimulus keeps rotating at the new position and the task is again, to identify the last orientation in the sequence. In 20% of the trials the stimulus stays unexpectedly in the centre.

This task is called the Jump-Task, because the best strategy for high scores is to make a saccade to the side, to which the stimulus was displaced. A correct re-fixation saccade allows to identify the orientation by foveal vision. Fig. 3.4 shows the results of the performance of the Jump-task of the same subjects in the second line. The result is almost the same as under the fixation condition of the original task. But interestingly the decrease of performance at higher ages is smaller. However, the analysis of the centre trials shown at the right of the figure show consistently lower scores than in the fixation task. This suggests that some of the younger subjects were prepared to move their eyes and presumably were thus no longer attending to the centre. Therefore they missed the correct orientation in a higher percentage of trials. Subjects of higher ages performed about as well as they did in the Fixation task.

A third version of the task (Antitask) was to replace the centre stimulus at some random time after the beginning of the trial by a distractor stimulus (a star symbol) at one of the two sides (randomly selected from trial to trial) and present the stimulus at the opposite side for a very short time in one of the 4 orientations. Again the task is to identify the orientation of the stimulus and press the corresponding key. As in the Jump task on 20% of the trials the centre stimulus was replaced by the target stimulus without presenting the distractor. In this task the best strategy for correct responses is to maintain fixation in the centre until the distractor occurs, to refrain from making a saccade to the distractor, and instead make a saccade to the target stimulus. With respect to the distractor such a saccade is called an antisaccade and therefore the task is called the Anti task.

Fig. 3.4 shows the age curves for this task in the third line. Clearly the scores are now considerably lower at young ages when compared with the Fixation task. If one looks at the centre trials the drop of scores is even more prominent. This suggests that the subjects were no longer attending to the centre and/or they were unable to suppress saccades to the distractor on some trials and therefore missed the target stimulus before it disappeared.

This section has shown that children at the beginning of school are already able to see pretty fast, when compared with the maximum reached at adult ages of 20 to 30 years. However, as in the case of low level auditory discrimination, the high performance level of the subjects in their best age is decreasing relatively early in life.

3.3. Subitizing and Counting by Memory

Subitizing is a special visual capacity, which rests not only on vision as a sensory process but also on short term visual memory. The word "subitizing" is derived from Latin "subito"

= fast, or at once. Subitizing means, that a given number of items can be recognized immediately without counting, even though they were seen only for a short period of time. We will see that subitizing in its restricted definition is possible only for item numbers of 3 or 4. For higher numbers one needs more time and a visual memory. Yet, we will use "subitizing" throughout for the sake of simplicity.

It has been discussed also whether the magno- and parvo-system contribute differentially to subitizing and counting [Simon et al. 1998] and therefore one might expect deficits in subitizing when other functions of the magno- or parvo-system exhibit anomalies.Subitizing became interesting in the context of cognition on the one hand and because of its possible role in basic arithmetic: it was argued, that children with specific deficits in dealing with numbers may have an incomplete sense of number. They know the digits (the visual signs of numbers) and the number of words (the auditory signs of numbers), but they may not have a stable inner presentation of how many items are associated with the digit and/or number of words. In the beginning, this was little more than a hypothesis. Now, systematic studies using standard methods of psychophysics are available, which support the notion of a relationship between subitizing and skills of basic arithmetic.

It is interesting to look at the history of numbers [Dehaene, 1999]. In many early cultures the digits for the numbers 1 to 3 or 4 were 1 to 3 points or 1 to 3 vertical or horizontal lines. The subjects could see how many items were meant by the form of the digits. The old Romans also used this way of writing small numbers. They did not invent new digits for higher numbers. They used the first letter of the word for the number to write the large numbers, like C for 100.

It was found that even infants have a possibility to differentiate between small numbers of objects [Starkey and Cooper, 1980]. It looks as if the recognition of small numbers of items is an essential visual capacity of humans and higher animals. The idea is, that this capacity is used to develop a sense of number. As a consequence there was an attempt to prove this hypothesis by examination of children, who had difficulties in basic arithmetic. A standard test procedure was designed and the data from normal control subjects were collected first, to build up the basis of a diagnostic procedure.

The Fig. 3.5 shows the sequence of frames in the task for examination of subitizing and counting by memory.

The items (small circles 2 mm in diameter) were presented for 100 ms on a LCD screen 6 cm x 2.5 cm in size. The screen was part of a hand held instrument, which also provided the digit keys from 0 to 9.

At least 12 practice trials with non-disappearing items were given. Subjects were informed that reaction time was important. Therefore they were also instructed to place their dominant hand just above the keyboard. After they understood the task all subjects were given another 5 trials of the real task for practice, i.e. presentation time was only 100 ms.

The central fixation point was presented first. After 1 sec the fixation point disappeared, the test pattern was presented, and the subject responded by pressing the response key. The next trial was initiated only after another key press by the subject. Each number of items was shown 20 times with the exception of a single item, which occurred only 10 times. Altogether, 170 trials were run for each subject. The total time for a test session was about 20 min or less.

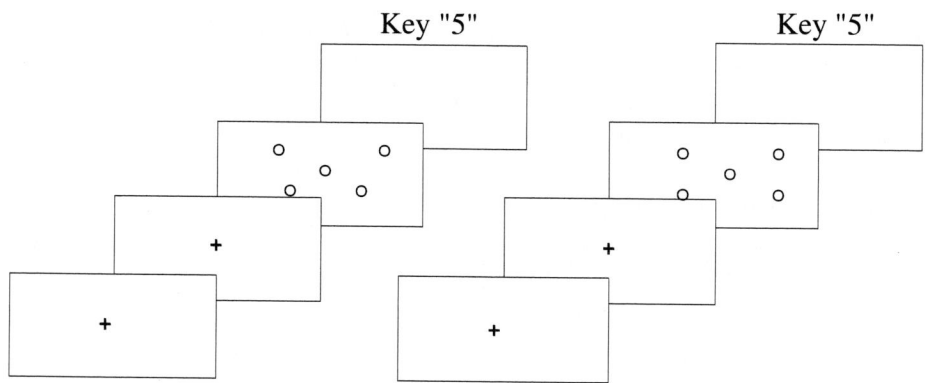

Figure 3.5. The figure shows two examples of sequences of frames in the task to examine subitizing and number counting. Time runs from left to right. The left part depicts a case, where the pattern of the 5 items is irregular, the right part shows the same number of items, but the pattern looks regular, as in the case of the number on a die.

First, we have to explain, how subjects solve the task in some detail. It is almost trivial that the tasks becomes more and more difficult as the number of items increase, while for small item numbers the correct response should be found easily and quickly. In fact, the Fig. 3.6 shows that for item numbers below 4 almost all responses are correct. Children (age 9 to 10), young adults (age 23 to 28), and older adults (age above 45) do not show significant differences. However, when the item numbers exceed 4, the percentage of correct responses decreases. Yet, 8 items can still be correctly seen in about 80% of the cases. Differences between children and adults become evident for item numbers above 4 or 5.

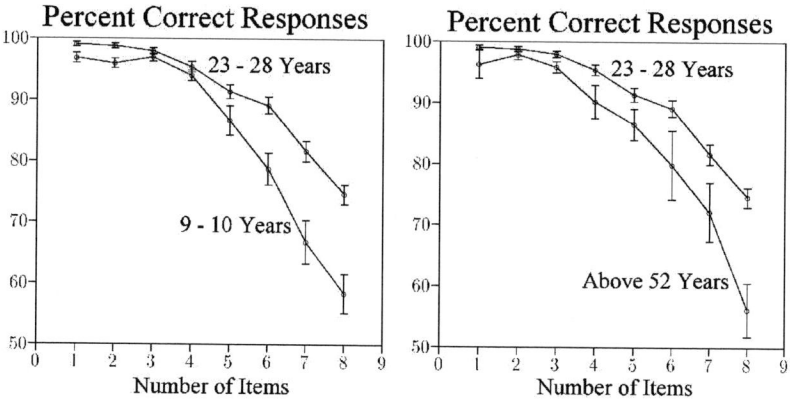

Figure 3.6. The percentage of correct responses as a function of the number of presented items. The left panel depicts the data from children and young adults, the right panel shows the data from young and older adults as indicated by the labels. Only small differences are obtained for small numbers of items. The curves diverge only for 4 items or more.

When looking at the response time of the correct responses (Fig. 3.7) one can see that 1 to 3 items are correctly seen after the same time. This is the reason, why the process of counting small numbers of items is called subitizing. With increasing item numbers the response times become longer in an almost linear fashion. This means, that each additional item takes about the same extra time. This seems to be the principle way of solving the task irrespective of the age of the subjects. Closer inspection of Fig. 3.7 shows that the overall response time and the slope of the linear regression line changes with age. While the percentage of correct responses was about the same for item numbers up to 3, the response times are different depending on the different age groups.

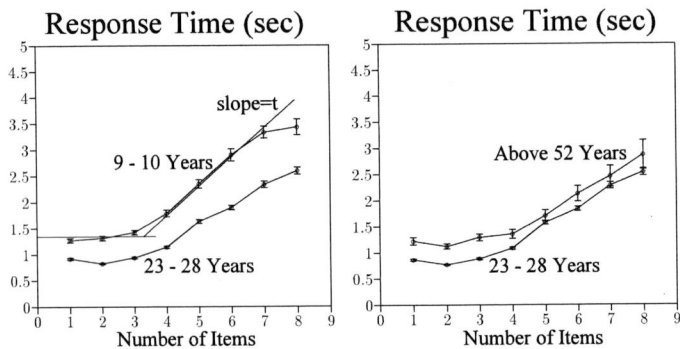

Figure 3.7. The response time of correct reponses as a function of item number. Note that for children and older adults the response times are longer even for small item numbers, while the percentages of correct responses are still close to 100%.

From these considerations we can define variables to describe the process of subitizing and number counting: the basic response time T is defined as the time for responding to the presentation of 1 item only. The slope of the linear regression line gives the time per item, tm. The mean value of correct responses is a measure of the correctness P of recognizing 4 to 8 items. We will see below, that while P decreases, tm increases. Therefore, we define an extra variable, which combines tm and P as the effective recognition speed ERS= P/tm.

The cross correlations of the variables show, that there are only two independent variables: one deals with time (T and tm show high correlations) and the other deals with correctness. Yet, we show the age development of all 4 variables.

The Fig. 3.8 shows the age curves of P and tm. The percentage of correct responses increases with age and reaches a maximum not before the age of 20 years. A moderate decline is seen for ages above 30 or 35 years. Similarly, the time per time decreases until ages above 20 years and a decline occurs for ages above 35 years.

The Fig. 3.9 shows the age curves for ERS and T. Both variables develop in about the same way as the other variables of the task: long lasting improvements with age and a decline above the age of 35 years. Only the response times reach their minimum already with the age of 15 years. These u-shaped age curves are similar to the curves describing the performance of number discrimination [Trick et al. 1996].

Since it was found that even babies have a basic capacity of subitizing and since we find, that even adults can subitize no more than 4 to 5 items, we have to pay special attention to

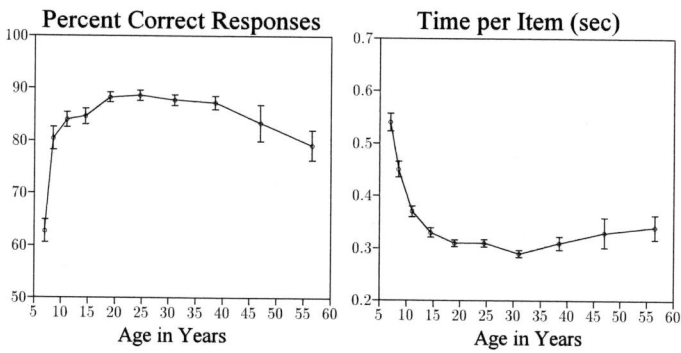

Figure 3.8. The age development of the percentage of correct responses and of the time per item.

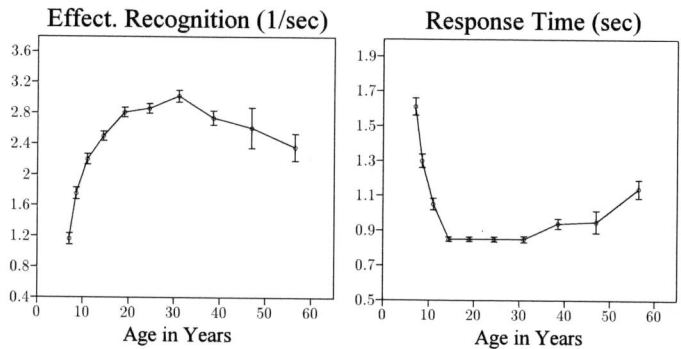

Figure 3.9. The age development of the effective recognition speed and the basic reaction time.

the variables that describe the performance for item numbers between 1 and 4. We calculate the mean response time for 1 to 4 items and the corresponding mean value of the correct responses. The latter should be close to 100%, because this is often used alone as the criterion for perfect (ideal) subitizing.

The Fig. 3.10 shows the two age curves. Now we see, that indeed almost perfect subitizing of 4 items is achieved at all ages except the youngest group given we use a 95% correct response as a criterion. However, the time to arrive at the correct number decreases with age. This means that perfect subitizing is possible, when considering the correct responses, but the capacity of subitizing still improves until the age of 20 years when considering the time. This fact will be reconsidered when we describe the diagnostic procedure to find developmental deficits and when we compare pre- and post-data of the training of subitizing.

In conclusion of this chapter we can say, that the basic visual functions described here do not reach the performance level of the adults before the age of 13 years (dynamic vision) or before the age of 18 years (subitizing). The visual functions have this long lasting development in common with the auditory functions described in the previous chapter. Of course,

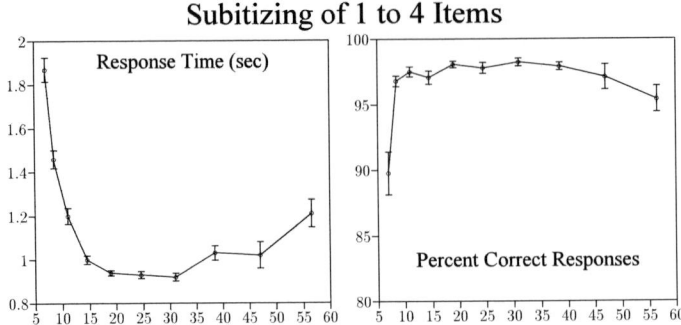

Figure 3.10. The age curves of subitizing for item numbers 1 to 4 are shown. Note that the response times change quite strongly with age, while the age curve of the percentage of correct responses is relatively flat.

one has to keep in mind, that there is a great deal of interindividual variation in single age groups, some of the members already performing pretty well and above the mean value, others way below the mean. These details can only be seen by looking at the distribution of the data.

We will use "subitizing" for the visual capacity of knowing the number of items between 1 and 9, even though only for item numbers 1 to 4 "subito" is correct, because for higher numbers more time is needed for the correct response (but the presentation time is short). In this sense the "subito" refers to the presentation time, not to the response time. As long as one deals with subjects of the same age, one misses the fact that subitizing is different in time for different ages, even though the ability to tell the correct number is about the same.

The role of the parietal cortex in number processing has been described earlier [Dehaene and Cohen, 1995]; [Dehaene et al. 1998]. We will see in the next part, that children with dyscalculia often suffer from deficits in subitizing and number counting and that a proportion of them suffer also from deficits in the frontal control of saccadic eye movements. Since the parietal and frontal cortical structures are intimately related this observation will not come as a surprise. Also, the coding of coding of magnitude is located, at least in part, in the prefrontal cortex [Dehaene, 1997].

Chapter 4

Development of Saccade Control

Summary

This chapter describes the development of eye movement control. We will consider, however, only those aspects of eye movements that are important for reading: stability of fixation and control of saccades (fast eye movements from one object of interest to another). The saccadic reflex and the control of saccades by voluntary conscious decision and their role in the optomotor cycle will be explained on the basis of the reaction times and neurophysiological evidence. The diagnostic methods used in the next part of the book will be explained in this chapter. The age curves of the different variables show that the development of the voluntary component of saccade control lasts until adulthood

4.1. Introduction

Saccades are fast eye movements. We make them all the time, 3 to 5 in a second. Due to these saccadic eye movements the brain receives 3 to 5 new pictures from the retina in each second. Without these ongoing sequences of saccades we would not see very much, because of the functional and anatomical structure of the retina: in the middle it contains a small area, called the fovea, where the receptor cells and the other cells in the retinal layers are densely packed. It is only this small part of the retina which allows us to see sharp images. Anything that we want to see in detail or what we want to identify as an object or any other small visual pattern, e.g. a letter, must be inspected by foveal vision. The solution for this biological demand are sequences of rapid relatively small eye movements (saccades). The time periods between saccades are 150 to 300 ms long. They are often called "fixations".

Usually in everyday life these saccades are made automatically, i.e. we do not have to "think" about them and we do not have to generate each of them by a conscious decision. But, by our own decision, we can also stop the sequence and fixate on a certain small object for longer periods of times. We can also actively and voluntarily move our eyes from one place of interest to another. This situation is quite similar to our breathing: it works by itself but we can control it also voluntarily.

We are not aware of these saccades, they remain unconscious and – most importantly – we do not see the jumps of the retinal image. Somehow, the visual system cooperates with the saccade control centres in a perfect way to differentiate between self-induced movements of the retinal image and externally generated movements.

Only under certain somewhat artificial conditions we can see our saccades. An example is shown by Fig. 4.1.

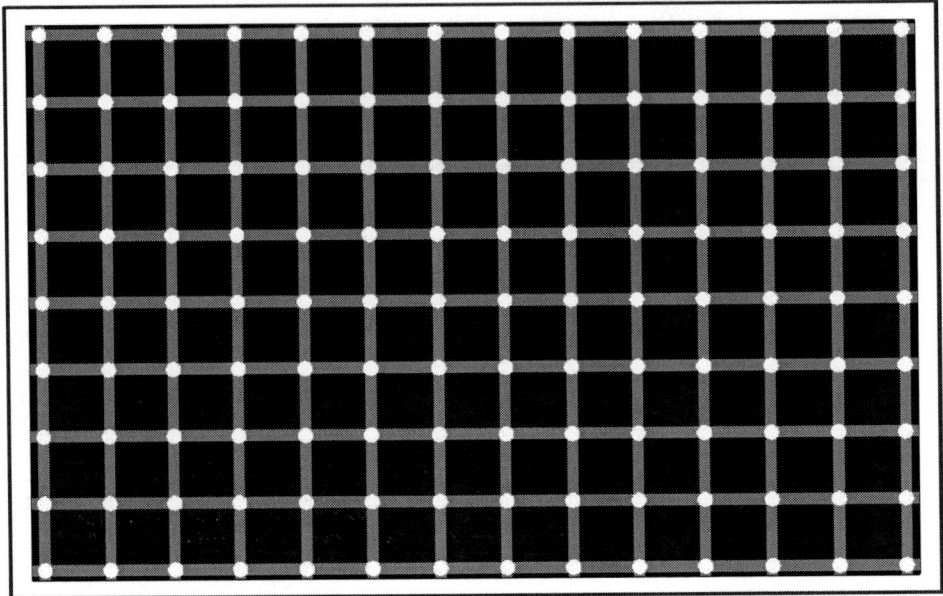

Figure 4.1. Scintillations of the Herman grid due to eye movements. As long as we look around across this pattern, we see black dots jumping around. Whenever we try to look at one of them, there is no black spot. When fixating the eyes at one white spot for a few seconds black spots are no longer seen. With still longer fixation most of the white dots disappear also.

The figure shows a modification of the well known Hermann grid. As long as we look around across on this figure, we see black dots appearing spontaneously at the crossing lines between the black squares. The picture looks like scintillations. We notice, that there are no such black dots. To avoid this illusory black blinks, we just have to decide to stop making saccades. The reader may try her/himself. Pick one of the white dots and maintain fixation at it. As long as one can prevent saccades, the black dots remain absent. Each saccade occurring spontaneously will create the illusion of dots again.

One can also see illusory movements, which are related to eye movements. The Fig. 4.2 shows an example. The movements disappear when we stop to make eye movements.

An example of a geometric illusion also allows one to become aware of one's own saccadic eye movements. The Fig. 4.3 shows the Z-illusion.

One can see that the prolongations of the short ends of the line (upper left and lower right) will not meet the corners at the lower left and upper right. If one succeeds to prevent all saccades for some seconds (up to 10, which is a long time in this context) one will see that the lines meet the corners as they do in reality. The reader may convince her/himself by using a ruler.

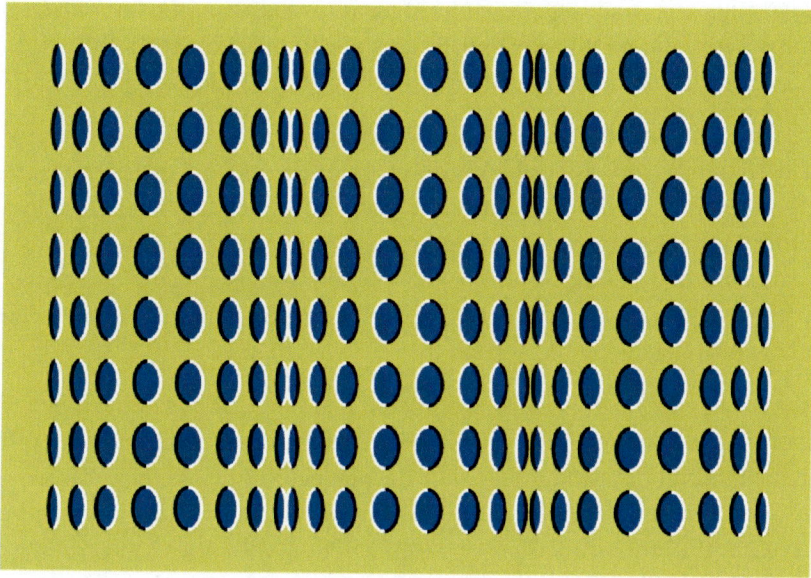

Figure 4.2. The horizontal movements that one sees when looking at this figure are obviously illusory, because nothing moves in this figure. The illusion disappears, when we stop our eye movements by fixating the centre of the figure.

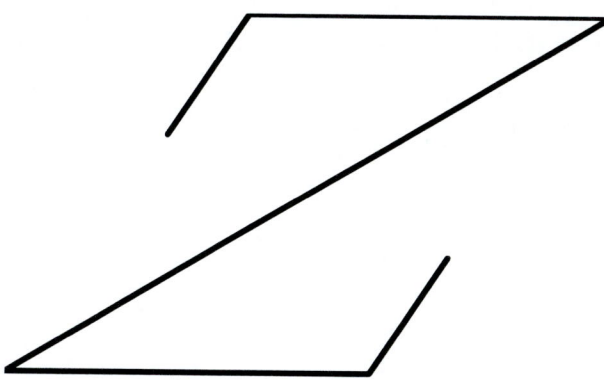

Figure 4.3. Z-Illusion shows the capital letter Z. The prolongations of the short lines do not seem to hit the corners. The real geometric relations can be seen by using a ruler and draw the prolongations or they can be seen by fixating the eyes in the middle.

As long as we do not take into account the fact that vision needs saccades, we will not understand the visual system. Any theory of vision which makes correct predictions but does not include the fast movements of the eyes can hardly be considered a valid theory.

It is interesting to note, that most visual psychophysical experiments require the subject to fixate one spot of light while being examined on their visual experience with another stimulus a distance away. The reason for this methodological detail is the sharply decreasing visual acuity from the centre to the periphery of the visual field. Unfortunately, when it came to visual illusions, fixation was not longer required when testing the stability of the illusory impression. The result was that the instability of geometrical illusions, shown in the examples above, remained undiscovered until recently [Fischer et al. 2003].

In particular, when we talk about reading, eye movements are one key for the under-standing of the reading process, which allows us to compose words from letters or from syllables. We therefore have to consider the saccade system, before we may consider its role in reading. The significance of saccadic eye movements in reading was emphasized also by a reader model resting on the basis of experiments, where eye movements were measured while the subjects were reading text, which was manipulated in several physical and linguistic ways [Reichle et al. 2003].

In the following sections we will consider those parts of eye movement control, that play an important role for reading. The other types of eye movements (vestibular ocular compensation, optokinetic nystagmus) will be neglected all together. However, we will consider the instability of fixation due to unwanted saccades or to unwanted movements of the two eyes in different directions or with different velocities. In principal, binocular vision is not needed at all for reading, but imperfections of binocular vision, which remain undetected, may disturb vision and consequently reading may be difficult.

The Fig. 4.4 shows the most important anatomical connections that begin in the retina and end up at the eye muscles.

Anatomical Pathways for Saccade Control

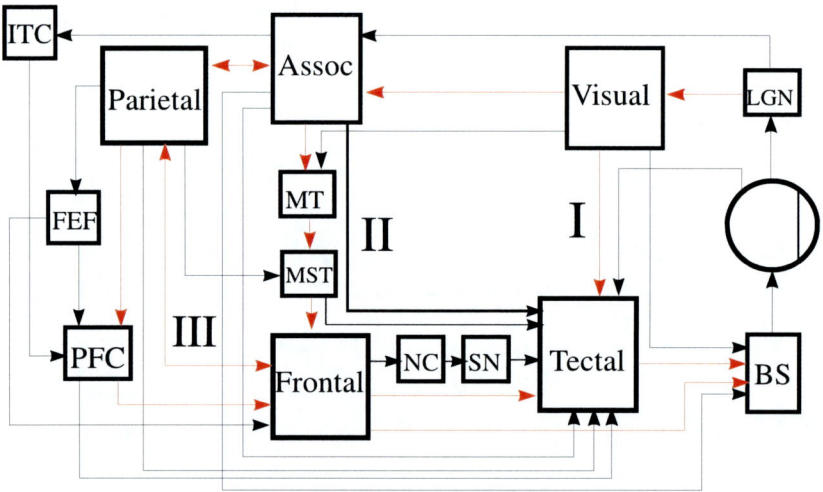

Figure 4.4. The figure shows a schematic diagram of the neural system of the control of visually guided saccades and their connections. LGN = Lateral Geniculate Nucleus; Assoc = Association Cortex; ITC = Infero-Temporal Cortex; FEF = Frontal Eye Field; PFC = Pre-frontal Cortex; MT = Medio Temporal Cortex; MST = Medio-Superior-Temporral Cortex; NC = Nucleus Caudatus; SN = Substantia Nigra; Tectal = Tectum = Superior Coilliculus; BS = Brain Stem.

There are hundreds of papers on eye movements in the literature. Saccades have always received special interest [Fischer, 1987]. Here we are interested in fixation and in saccades. Today it is easy to measure eye movements in human subjects. A sufficiently precise method uses infrared light reflection from the eyes. The data presented here are all collected by using this method. For the purpose of clinical application a special instrument was developed. It is called ExpressEye and provides the infrared light source, the light sensitive elements, the visual stimuli needed for basic tasks (with minilasers, see below), and the amplifiers. The instrument can deliver the raw eye position data trial by trial during the experiment and detect saccades and provides a statistical analysis of saccades. The front view of the system is shown in Fig. 4.5. The method has been described in detail elsewhere [Hartnegg and Fischer, 2002]. The raw data can be stored at the hard disc of a computer for further analysis to obtain different variables that characterize the performance of the tasks. These methods have been described in detail [Fischer et al. 1997].

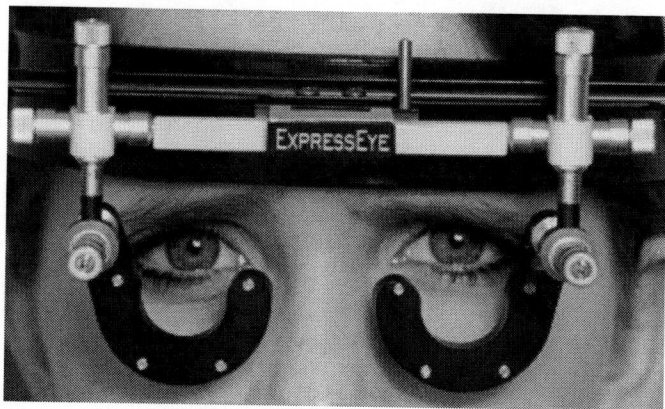

Figure 4.5. The front view of the Express Eye designed to measure eye movements. One sees the screws for the mechanical adjustment in all 3 dimensions in front of each eye. An infrared light emitting diode and the two photocells are located behind and directed to the centre of the eye ball.

4.2. Fixation and Fixation Stability

It may come as a surprise, that a section on eye movements starts by dealing with fixation, i. e. with periods of no eye movements. It has been the problem over many years of eye movement research, that fixation was not considered at all as an important active function. The interest was in the moving and not in the resting (fixating) eye. Only direct neurophysiological experiments [Munoz and Wurtz, 1992] and thorough investigation of the reaction times of saccades [Mayfrank et al. 1986] provided the evidence, that fixation and saccade generation are controlled in a mutual antagonistic way similar to the control of other body muscles. We will see, that we can observe movements of the eyes during periods where they were not supposed to move at all. It seems that there was little doubt, that almost any subject can follow the instruction "fixate" or "do not move the eyes". But this not the case:

stability of fixation cannot always be guaranteed by all subjects. This section deals with the results of the corresponding analysis of eye movements.

4.2.1. Monocular Instability

As pointed out earlier, we have to consider two different aspects of disturbances of fixation: the first aspect are unwanted (or intrusive) saccades. These are mostly small conjugate saccades that take the fovea from the fixation point and back. This kind of disturbance is called a monocular instability, because when it occurs one sees it in both eyes at exactly the same time and by the same amount of saccade size. The disturbance remains if one closes one eye and therefore it disturbs monocular vision and it does not disturb binocular vision. This is the reason why it is called a monocular instability. Below we will also explain the binocular instability.

To measure the stability or instability of fixation due to unwanted saccades, we simply count these saccades during a short time period, when the subject is instructed to fixate a small fixation point. Such a period repeatedly occurs in both diagnostic tasks that are used for saccade analysis that are described in section 4.3.2. on page 44.

The number of unwanted saccades counted during this task is used as a measure of monocular fixation instability. For each trial this number is recorded and attributed to this trial. The mean value calculated over all trials will serve as a measure. The ideal value is zero for each individual trial and therefore the ideally fixating subject will receive also zero as a mean value.

The Fig. 4.6 shows the mean values of the number of intrusive (unwanted) saccades per trial as a function of age. While children at the age of 7 produce one intrusive saccade every 2 or 3 trials, adults around 25 years of age produce one intrusive saccade every 10 trials. At higher ages the number of intrusive saccades increases again. Of course, not every intrusive saccade leads to an interruption of vision and therefore one can live with a number of them without problems. But if the number of intrusive saccades is too high, visual problems may occur.

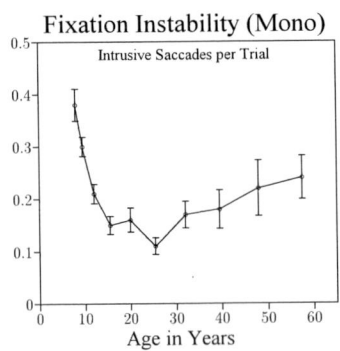

Figure 4.6. The curve shows the age development of the number of unwanted (intrusive) saccades per trial. The ideal value would be zero.

When we measure the monocular instability by detecting unwanted saccades, we should not forget, that there may be also another aspect of strength or weakness of fixation, which cannot be detected by looking at the movements of the eyes during periods of fixation, but rather be looking at reaction times of saccades that were required when the subject has to disengage from a visible fixation point.

4.2.2. Binocular Instability

To understand binocular stability we have to remember that the two eyes must be in register in order for the brain to "see" only one image, even though each eye delivers its own image. This kind of convergence of the lines of sight of the two eyes at one object is achieved by the oculomotor system. We call it motor fusion. However, even with ideal motor fusion, the two images of the two eyes will be different, because they look at a single three dimensional object from slightly different angles. The process of perceiving only one object in its three dimensions (stereo vision), is called perceptual fusion, or stereopsis.

When we talk about stereo vision (stereopsis) we mean fine stereopsis, i.e. single three-dimensional vision of objects. It is clear that we need both eyes for this kind of stereopsis. However, we also have three-dimensional vision with one eye only. The famous Necker cube shown in Fig. 4.7 is one of the best known examples. From the simple line drawing our brain constructs a three-dimensional object. Close one eye and the percept of the cube does not change at all. This type of three-dimensional spatial vision does not need both eyes. The brain constructs a three-dimensional space within which we see objects.

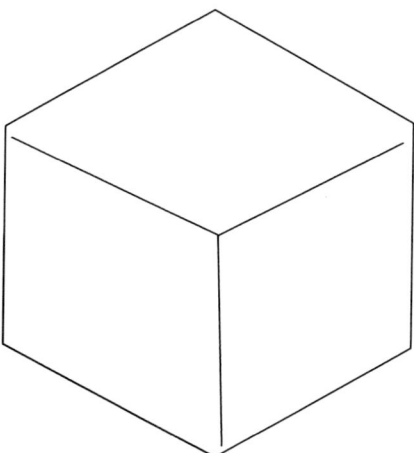

Figure 4.7. The figure shows the famous Necker cube. Note that one sees a three-dimensional object even though the lines are not correctly connected. The percept of a three-dimensionl cube remains even when we close one eye. Note, that the lines do not really meet at the corners of the cube. Yet, our perception is stable against such disturbances and the impression of a cube is maintained.

In order to guarantee stable stereopsis, the two eyes must be brought in register and they have to stay in register for some time. This means that the eyes are not supposed to move independently from each other during a period of fixation of a small light spot. By recording the movements of both eyes simultaneously one has a chance to test the quality of the stability of the motor aspect of binocular vision.

The Fig. 4.8 illustrates the methods for determining an index of binocular stability.

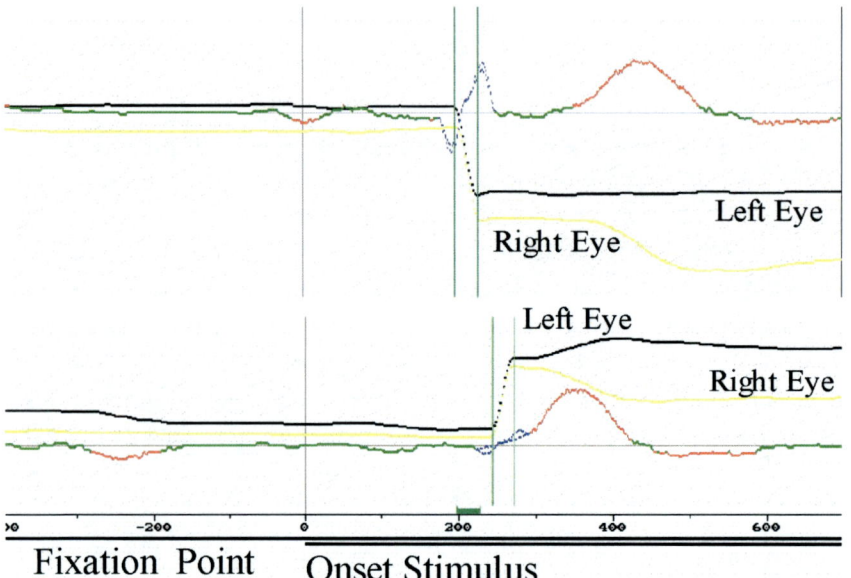

Fixation Point   Onset Stimulus

Figure 4.8. The figure illustrates the methods for determining an index of binocular stability by analysing the relative velocity of the two eyes. Time runs horizontally. At the time of stimulus onset the subject was required to make a saccade. Up means right, down means left. Two trials from the same child are depicted. In the upper case the left eye shows stable fixation before and after the saccade. The right eye, however, converges after the saccade producing a period of non-zero relative velocity. In the lower case, both eyes produce instability after the saccades. For details see text.

Two trials from the same child are depicted. In the upper trial the left eye shows stable fixation before and after the saccade. The right eye, however, converges after the saccade producing a period of non-zero relative velocity. In the lower case, both eyes produce instability after the saccades. The example shows, that the instability is sometimes produced by one eye only, or by both eyes simultaneously. Often it is caused in some trials by one eye, and in other trials by the other eye. Extreme dominance of one eye producing the instability was rarely seen (see below).

In the example of Fig. 4.8 the index of binocular instability was 22%. This means, that the two eyes were moving at different velocities during 22% of the analysed time frame. To characterize a subject's binocular stability as a whole, the percent number of trials, in which this index exceeded 15% was used. The ideal observer will be assigned zero. The worst case would be assigned a value of 100%.

The Fig. 4.9 shows the data of binocular instability of a single subject. The upper left panel depicts the frequency of occurrence of the percentages of time, during which the eyes were not in register. The upper right panel depicts the distribution of the relative velocity of the two eyes. The scatter plot in the lower left panel displays the correlation between these variables. Ideally all data points should fall in the neighbourhood of zero. The lower right panel depicts the time development of the variable percent time of limits by showing the single values as they were obtained trial by trial from trial 1 to trial 200. This panel allows us to see whether or not fatigue has in influence on the binocular stability.

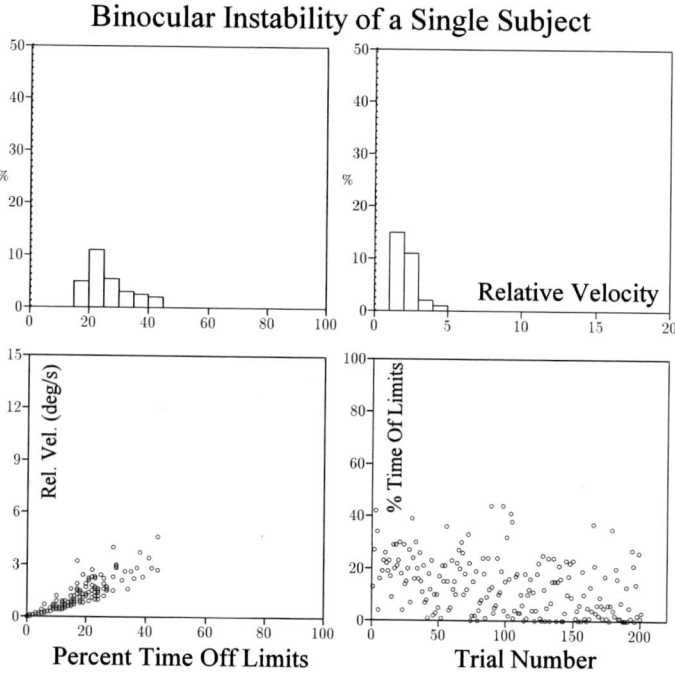

Figure 4.9. The figure shows the data of binocular instability of a single subject. For details see text.

When the values of the binocular stability were compared among each other, several aspects became evident: (i) Within a single subject the values assigned to the trials can be very different. Almost perfect trials may be followed by trials with long periods of instability. This means, that the subject was not completely unable to maintain the line of gaze for both eyes, but from time to time the eyes drifted against each other. (ii) There was a large interindividual scatter of the mean values even within a single age group. (iii) Even among the adult subjects large amounts of instability were observed. (iv) The test-retest reliability was reduced by effects of fatigue or general awareness of the subjects.

The Fig. 4.10 shows the age development of the binocular instability using data from the prosaccade task with overlap conditions.

At the beginning of school large values but also small values were obtained. There was a clear tendency towards smaller values until the adult age. However, the ideal value of zero is not reached at any age. This means that somehow small slow movements of the two eyes in different directions during short periods of time are well tolerated by the visual system. In other words: there are subjects with considerably instable binocular fusion not

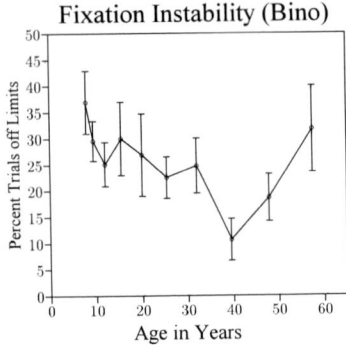

Figure 4.10. The percentage of trials in which the two eyes were moving relative to each other (in more than 15% of the analysis time) is shown as a function of age.

complaining about vision problems. Maybe these subjects suppress the "picture" of one eye to avoid double vision all the time at the price of a loss of fine stereo vision, their vision is monocular. Because this does not create too much of a problem in everyday life, subjects do not show up in the eye doctors' praxis. Their binocular system is never checked. This situation could be regarded as similar to the case of red-green colour blindness, which may remain undetected throughout life, because the subject has no reason to take tests of colour vision.

4.2.3. Eye Dominance in Binocular Instability

Since the two eyes send two different pictures to the brain, the picture of one of the two eyes must be prevented from automatically reaching consciousness. This is true for most of the visual scene we see. Only those parts form one single picture in the brain that fall on corresponding points of the two retinae. It is often forgotten, that this part covers only those objects, that are at about the same distance from the eyes as the object, that we are just fixating with both eyes.

Because of the necessity to suppress the information from one eye most of the time, it has been speculated, that each subject selects one eye as the dominant eye (similar to the selection of one hand as the dominant hand). If, however, the image of one eye is permanently suppressed, fine stereopsis is not possible. We can easily see, that the images of both eyes are present in our visual system, but usually, we do not perceive both of them. Fixating a near point and attending to an object further away leads to double vision of the object. The reader may try by her/himself using the thumb of one hand as a near point and the thumb of the other hand as a far point.

The Fig. 4.11 shows the distribution of the differences between the right eye values and the left eye values of the index of binocular instability. The mean value is not significantly different from zero. But one sees that in a few cases clear dominances are obtained (8 subjects scored values far to the left, 3 far to the right).

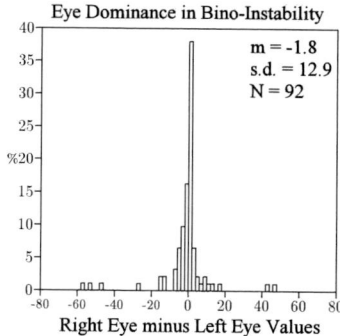

Figure 4.11. The distribution of the differences between the right eye and the left eye values of binocular instability.

4.2.4. Independence of Mono- and Bino-Fixation Instability

In principle, the two types of instability may have the same reason: a weak fixation system allows all kinds of unwanted eye movement, including unwanted saccades and unwanted drifts of one or both eyes. In this case one should see high correlations between the two variables describing these types of instability. Because both variables depend on age, we analyse the data within the restricted age group.

The Fig. 4.12 shows the scatter plots of the binocular versus the monocular instability for two age groups. The correlation coefficients were only 0.22 for the younger subjects (left side) and 0.21 for the older group (right side). Both correlations failed to reach a significance level of 1%. This means that the properties assessed by the two measures of fixation instability are independent from each other and are different in nature. When we look at the data of dyslexic children, we will have many more data that will support the independence of these two aspects of fixation instability. Also, we will see later, that monocular training improves the binocular instability but not the monocular instability.

4.3. Development of Saccade Control

In the last section we have considered the stability of fixation as an important condition for perfect vision. Earlier, we have mentioned, that saccades are necessary for vision. This might sound like a contradiction. The real requirement is, that one should be able to generate sequences of saccades and fixations without an intrusion of unwanted saccades and without losing the convergence of the two eyes when they are in register for a given distance. Therefore, both components of gaze control should function normally. This section will show that saccade control has to be subdivided into subfunctions described by different variables. We have to find out first, what these subfunctions are and how they can be assessed.

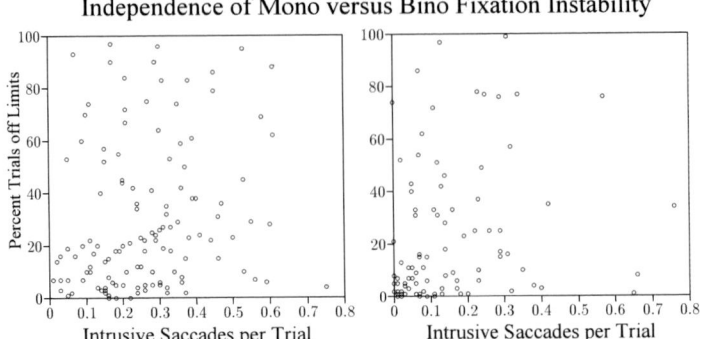

Figure 4.12. Scatter plot of the binocular versus the monocular instability obtained from overlap prosaccade trials. The left panels depict the data from subjects between 7 and 17 years of age (N= 129), the right panel shows the data from older subjects 22 to 45 years of age (N=97). No correlations can be seen.

4.3.1. The Optomotor Cycle and the Components of Saccade Control

The control of saccades has been investigated for about 40 years and still we do not understand the system completely. Specialization of visual scientists and oculomotorists has prevented the two fields so closely related from having been investigated by corresponding combined research projects for a long time. The oculomotor research groups were interested in the eye movement as a movement. Their interest begins when the eyes begin to move and it stops when the eyes stop to move. The visual groups on the other hand, were interested in time periods when the eyes do not move. They required their subjects to fixate on a small spot, while being tested for their visual functions. Only when the interest was concentrated on the time just before the movements, there was a chance to learn more about the coordination of saccades and visual processes.

The time before a saccade is easily defined by the reaction time: one asks a subject to maintain fixation at one spot of light straight ahead and to make a fast eye movement to an other spot of light, as soon as it appeared. Under these conditions the reaction time is in the order of 200 ms. This is a value, which one can find in a student handbook.

However, there were several problems. The first was: why is this time so long? While this was a question from the beginning there was no answer until 1983/84, when the express saccade was discovered in monkeys [Fischer and Boch, 1983] and in human observers [Fischer and Ramsperger, 1984]. The express saccades is the reflex movement to a suddenly presented light stimulus after an extremely short reaction time (70-80 ms in monkeys and 100-120 ms in human observers). The reflex needs an intact superior colliculus [Schiller et al. 1987].

The Fig. 4.17 shows in its lower part a distribution of saccadic reaction times. It exhibits 3 modes: one at about 100 ms, the next at about 150 ms, and the third at about 200 ms. It was evident from these observations, that there was not only one reaction time with a (large and unexplained) scatter. Rather the reaction time spectrum indicated that there must be at least 3 different presaccadic processes that determine the beginning of a saccade, each taking its own time in a serial way. Depending on how many of the presaccadic processes are already completed before the occurrence of the target stimulus, the reaction time can take one out of three values, each with a certain amount of scatter [Fischer et al. 1995].

It became clear that the shortest reaction time was 100 ms (not 200 ms) and this was much easier to explain by the time the nerve impulses needed to be generated in the retina (20 ms), to travel to the cortex (10 ms), to the centres in the brain stem (10 ms) and finally to the eye muscles (15 ms). Another 5 ms elapse before the eye begins to move. A much shorter time remains, that was attributed to a central computation time to find the correct size of the saccade to be programmed. One has to know at this point, that saccades are pre-programmed movements: during the last 80 ms before a saccade actually starts, one cannot change anything anymore.

The next problem was: what is it that keeps the eyes from producing saccades all the time? Or, the other way around: what is it that enables us to fixate on an object on purpose? The answer came from observations of cells that were active during time periods of no eye movements and that were inhibited, when saccades were made [Munoz and Wurtz, 1993]. What could have been found much earlier, became clear only after the neuroscientists began to think in very small steps: each process, that we experience as one unique action, must be eventually subdivided into a number of sub-processes. It became clear that the break of fixation and/or allocated attention was a necessary step before a saccade can be generated. There is quite a number of single papers contributing to the solution of the related problems. Most of them have been summarized and discussed earlier [Fischer and Weber, 1993]. Most important for the understanding of the relation between saccades and cognitive processes is the finding that there is a component in saccade control that relies on an intact frontal lobe [Guitton et al. 1985]

From all these considerations it became clear, that sequences of fixations and reflexes form the basis of natural vision.

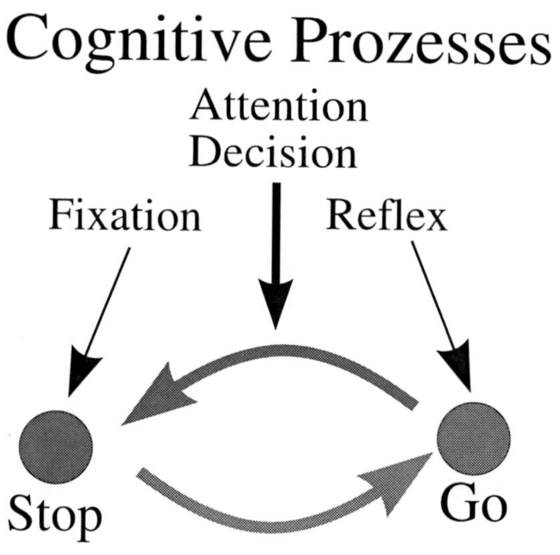

Cognitive Prozesses
Attention
Decision

Fixation   Reflex

Stop   Go

Figure 4.13. The figure shows the functional principle of the cooperation of the 3 components of eye movement control.

The Fig. 4.13 shows a scheme, which summarizes and takes into account the different findings: the stop-function by fixation alternates with the reflex, the go-function. These two together built up a stop-and-go traffic of fixations and saccades. What remains open, was the question of how it was possible to interrupt this automatic cycling. The answer came from observations of the frontal lobe functions: patients who lost parts of their frontal lobe on one side, were unable to suppress the reflex in a simple tasks, called the antisaccade task [Guitton et al. 1985]. This task requires the subject to make a saccade to one side, when the stimulus is presented at the opposite side. The task became very popular during recent years, but it was used already many years ago [Hallet, 1978].

4.3.2. Methods and Definition of Variables

The fundamental aspects of saccade control as described by the optomotor cycle have been discovered by using two fundamental tasks, which are surprisingly similar but give insight into different aspects of the optomotor cycle. They have been used to quantitatively measure the state of the system of saccade control. We describe these methods and define the variables first. Then we will see some of the results obtained by the these methods.

The two tasks are called the overlap prosaccade task and the gap antisaccade task. The words pro and anti in their names refer to the instructions that the subject is given. The words overlap and gap describe the timing of the presentation of the fixation point. The Fig. 4.14 shows the sequence of frames for gap and for overlap conditions.

In both tasks a small light stimulus is shown, which the subjects is asked to fixate. This stimulus is called the fixation point.

In overlap trials a new stimulus is added to the left or right of the fixation point. The subjects are asked to make a saccade to this new stimulus, the target stimulus, as soon as it appears. Both, the fixation point and the target are visible throughout the rest of the trial: they overlap in time. This overlap condition and the task to look towards ('pro') the stimulus explain the complete name of the task: overlap prosaccade task.

The gap condition differs from the overlap condition in only one aspect: the fixation point is extinguished 200 ms before the target stimulus is presented. The time span from extinguishing the fixation point to the onset of the new target stimulus is called gap. In addition to this physical difference the instruction for the subject is also changed: the subject is required to make a saccade in the direction opposite ('anti') of the stimulus: when the stimulus appears left, the subject shall look to the right and vice versa. Therefore the complete name of this task is: gap antisaccade task.

The prosaccade task with overlap condition allows us to find too slow or too fast reaction times and to measure their scatter. The presence of a fixation point should prevent the occurrence of too many reflexes, the appearance of a new stimulus should allow a timely generation of a saccadic eye movement.

The antisaccade task with gap condition challenges the fixation system to maintain fixation and the ability to generate a saccade against the direction of the reflex.

Now we can define variables to describe the state of the saccade control system. First of all, one has to keep in mind, that these variables may be different for left versus right stimulation. Because left/right directed saccades are generated by the right/left hemisphere, side differences should not be much of a surprise. However, for the general definition of the

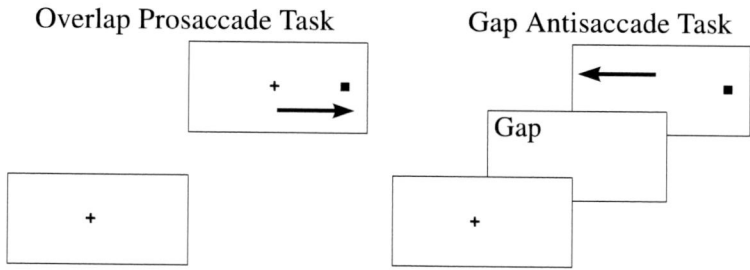

Figure 4.14. The figure shows the sequence of frames for overlap and for gap conditions. The horizontal arrows indicate, in which direction the saccade should be made: to the stimulus in the prosaccade task, in the direction opposite to the stimulus, in the antisaccade task.

variables to be used in the diagnosis, the side differences do not need to be considered at this point. The Fig. 4.15 illustrates the definition of the variables described below. Time runs from left to right. The stimulus is indicated by the thick black line. Because its presentation is identical in both conditions, it is drawn only once in the middle. The fixation point is shown by the thin black line. In the case of an overlap trial, the fixation remains visible, in the case of a gap trial the fixation point is extinguished 200 ms before. In addition the figure shows schematical traces of eye movements, which help to understand the definition of the variables. The upper case shows a trace from an overlap trial. Usually one saccade is made and it contributes its reaction time, SRT. Below one sees two examples of traces. One shows a correct antisaccade, which contributes its reaction time, ANTI-SRT. The other trace depicts a trial with a direction error that was corrected a little later. It contributes the reaction time of the error, Pro-SRT and the correction time, CRT (in case the error was corrected). While these variables can be taken from every single trail, some other variables are determined by the analysis of the complete set of 200 trials: the percentage of express saccades from all overlap trials, the percentage of errors from all gap trials and the percentage of corrections among the errors.

List of variables:

From the overlap prosaccade task the following mean values and the scatter are used:

- SRT: saccadic reaction time in ms from the onset of the target to the beginning of the saccade

- % expr: the percentage of express saccades, i.e. reaction times between 80 and 130 ms.

From the gap antisaccade task:

- A-SRT: the reaction time of the correct antisaccades

- Pro-SRT: the reaction time of the errors

- CRT: the correction time

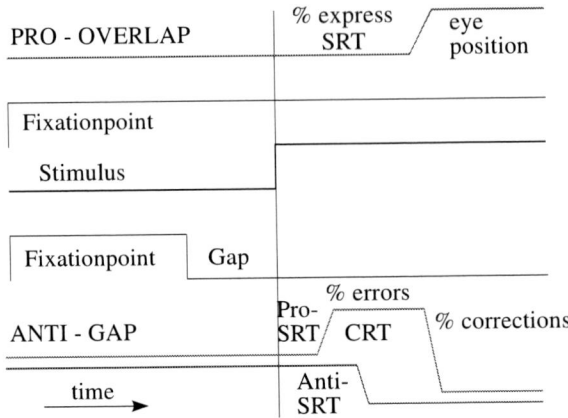

Figure 4.15. The schematic drawing of eye movement traces illustrates the definition of the different variables describing the performance of the prosaccade task with overlap conditions and the antisaccade task with gap conditions.

- %err: the percentage of errors

- %corr: the percentage of corrections among the errors

Note that the percentage of trials, in which the subject missed to reach the opposite side within the time limit in the trial (700 ms from stimulus presentation) can be calculated as

$$= \quad {}^a\!A\!E\,(100 - \quad \pm A\!E)/100$$

The latter variable combines errors rate and correction rate.

4.3.3. Prosaccades and Reflexes

The optomotor system has a number of reflexes for automatic reactions to different physical stimulations. The best known reflex is the vestibular-ocular reflex, which compensates head or body movements to stabilize the direction of gaze on a fixated object: the eyes move smoothly in the direction opposite to the head movement in order to keep the currently fixated object in the fovea. Similarly, is it possible to stabilize the image of a moving object by the optokinetic reflex. Both reflexes have little or nothing to do with reading.

The saccadic reflex is a reaction of the eyes to a suddenly appearing light stimulus. It was discovered only in 1983/84 by analysing the reaction times of the saccades in a situation, where the fixation point was extinguished shortly (200 ms gap) before a new target stimulus was presented. It was known at that time that under these gap conditions the reaction times were considerably shorter as compared to those obtained under overlap conditions [Saslow, 1967]. When the gap experiment of Saslow was repeated years later, it became evident that among the well known reactions around 150 ms after target onset there was a separate group of extremely short reactions at about 100 ms, the express saccade [Fischer and Ramsperger, 1984].

The Fig. 4.16 shows the distribution of reaction times from a single subject. One clearly sees two peaks. The first peak consists of express saccades, the second represents the fast regular saccades.

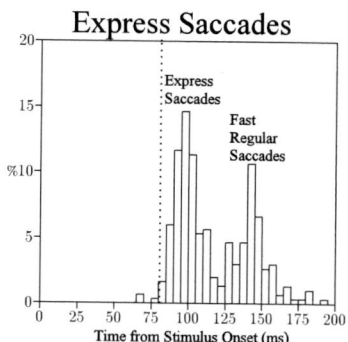

Figure 4.16. The figure shows the distributions of reaction times from a single subject. One clearly sees two peaks. The first represents the express saccades, the second the fast regular saccades.

The Fig. 4.17 shows the difference in the distributions of reaction times when gap and overlap trials were used. The separate peaks in the distributions indicate, that saccades can be generated at distinctly different reaction times depending on the preparatory processes between target onset and the beginning of the saccade. In the gap condition there is time during the gap to complete one or even two pre-saccadic processes. Therefore the chances of generation of express saccades is high. In the overlap condition it is the target stimulus, which triggers the preparatory processes and therefore the chances of express saccade are low.

If one leaves the fixation point visible throughout the trial (overlap condition) the reaction times are considerably longer, even longer as compared with the gap=0 condition (not shown here).

The consistent shortening of reaction time by introducing a temporal gap between fixation point offset and target onset was surprising, because the role of fixation and of the fixation point as a visual stimulus was unknown. The effect is called the gap-effect and has been investigated in numerous studies of different research groups all around the world since 1984. The effect of the gap on the reaction time is strongest if the gap lasts approximately 200 Milliseconds. An overview and a list of publications can be found in an overview article [Fischer and Weber, 1993].

Today it is clear, that the main reason for the increase in reaction time under overlap conditions is due to an inhibitory actions of a separate subsystem in the control of eye movements, the fixation system. It is activated by a foveal stimulus, which is being used as a fixation point and it inhibits the subsystem which generates saccades. If this stimulus is removed early enough, the inhibition is removed by the time the target occurs and a saccade can be generated immediately, i.e. after the shortest possible reaction time.

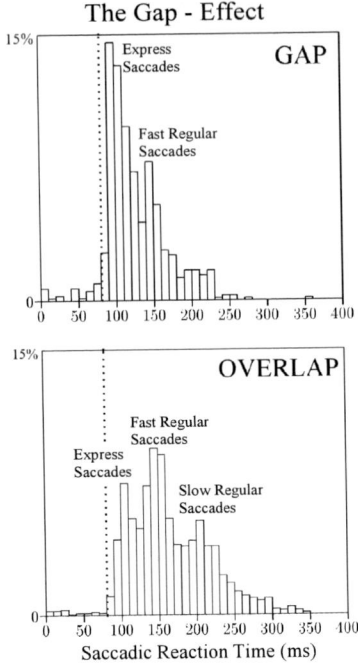

Figure 4.17. The figure shows the difference in the distributions of reaction times when gap and overlap trials were used. Note the separate peaks in the distributions.

Note that the effect of the gap is not a general reduction of reaction times, but rather the first peak is larger and the third peak is smaller or almost absent. As a result the mean value of the total distribution is reduced.

At this point, we do not have to go through the discussion of whether or not directed visual attention also inhibits or uninhibits the saccade system depending on whether attention is engaged or disengaged. This issue has been discussed extensively and still today the different arguments are not finally settled. We only have to keep in mind, that the gap conditions enables the reflex movements to a suddenly presented visual stimulus.

It is also important to remember, that there are subjects, who produce express saccades under overlap conditions [Fischer et al. 1993]. We will encounter these so called express saccade makers [Biscaldi et al. 1996] again, when we consider the eye movements of dyslexic subjects.

4.3.4. Antisaccades: Voluntary Saccade Control

It is an everyday experience, that we can stop our saccades and that we can direct our centre of gaze to a selected object or location in space on our own decision. These saccades are called voluntary saccades for obvious reasons. All from the beginning it will not be a big surprise to learn, that there are also different neural subsystems, that generate the automatic saccades and the voluntary saccades.

The investigation of voluntary saccades was introduced many years ago [Hallett, 1978], but the oculomotor research community did not pay attention to it very much. Hallett instructed his subjects to make saccades to the side opposite of a suddenly presented stimulus. These saccades were called antisaccades.

An early observation of neurologists did not receive much attention either, but turned out to be very important. It was reported that patients, who lost a considerable part of their frontal lobe in one side only, were unable to generate antisaccades to the side of the lesion, while the generation of antisaccades to the opposite side remained intact [Guitton et al. 1985]. Meanwhile the antisaccade task has become an almost popular "instrument" for diagnosis in neurology and neuropsychology. Reviews have been published and can be consulted by the interested reader [Everling and Fischer, 1998]; [Munoz and Everling, 2004].

The effect of changing the instruction from " look to the stimulus, when it appears" to "look away from from the stimulus (the anti-instruction)" can be seen in Fig. 4.18.

Reaction Times of Antisaccades

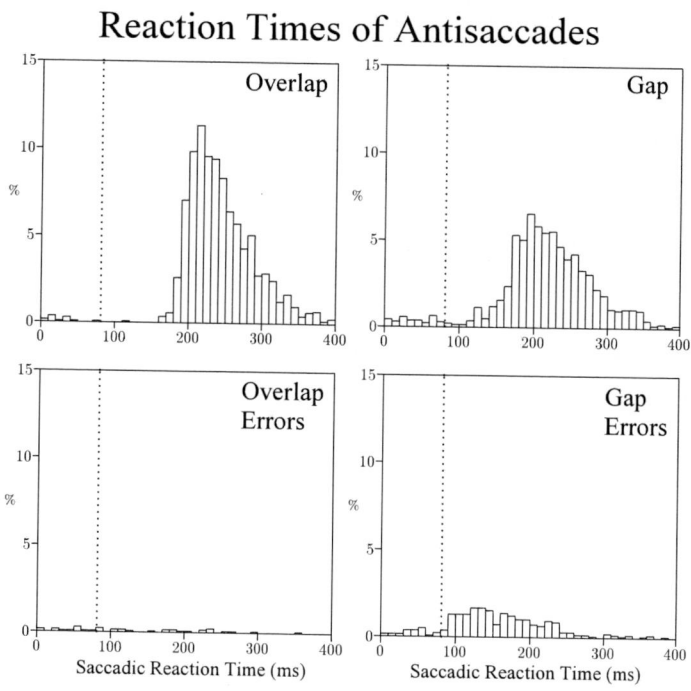

Figure 4.18. The figure shows the distribution of saccadic reaction times under overlap (left) and under gap conditions (right). The lower panels show the data from those trials, in which the subjects made errors by looking first to the stimulus. Note, that with overlap conditions there are virtually no such errors, while with gap conditions a considerable number of errors were made.

The introduction of the gap leads to quite a number of errors. Interestingly, subjects often failed to judge their performance: some claimed that they made many errors, but did not make many, others claimed that they made few errors, but made quite a lot. This indicates that we have little conscious knowledge of what we do with our eyes. The processes preparing the saccades and their execution remain mostly unconscious. When the variables obtained from an overlap prosaccade task and from a gap antisaccade task were analysed by a factor analysis [Gezeck et al. 1997] it turned out, that there were only 2 factors. The first factor contained the variables that describe prosaccades, irrespective of whether they were generated in the prosaccade task or as erros in the antisaccade task. The second factor contained the variables that described the performance of the antisaccade task. But there was one exception: the error rate loaded on both factors. The explanation of this result becomes evident, when we remember that the correct performance of the antisaccade task requires 2 steps: suppression of the prosaccades and generation of antisaccades. The error may be high for 2 reasons: (i) the suppression is not strong enough, (ii) the subject has difficulties in looking to the side where there is no target.

The details of these observations finally resulted in the decision to use the 2 tasks described above in order to characterize the functional state of the system of saccade control. The procedure of the corresponding analysis of the raw eye movement data have been described in great detail [Fischer et al. 1997]. The tasks are illustrated in Fig. 4.14, the definition of the variables are illustrated by Fig. 4.15. Today, there is already a special instrument and analysis system, which allows us to measure the eye movements and to assess the variables, their mean values and their scatter. Test-retest reliability of saccade measures, especially for measures of antisaccade task performance are available [Klein and Fischer, 2005]

The data obtained from these two tasks are shown in Fig. 4.19 separately for left and right stimulation. The data were combined from 8 subjects in the age range of 14 to 17 years. The distributions show most of the important aspects of the data, which are not as clear in the data of single subjects. The 3 peaks are seen in both left and right distributions obtained from the prosaccade task with overlap conditions (upper panels). But they are not quite identical: more express saccades are made to the right stimulus than to the left stimulus. The antisaccades (lower panels) have longer reaction times and a structure with different modes is missing.

Earlier studies of antisaccade performance as summarized recently [Munoz and Everling, 2004] analyse the reaction times in the antisaccade task and the percentage of errors. Most studies, however, failed to analyse the reaction time of the errors. They also neglected the percentage of corrective saccades and the correction time. We will see below that these variables provide important information about the reasons why errors were made [Fischer et al. 2000]; [Fischer and Weber, 1992].

We therefore also show the distributions of the reaction times of the errors and the distributions of the correction times of the same subjects as in Fig. 4.19.

Now we can further explain the data shown in Fig. 4.19 and in Fig. 4.20. The subjects as a group made quite a number of express saccades in the overlap prosaccade task (upper panels of Fig. 4.19) indicating that their ability to suppress saccades is limited. We can see that the errors in the gap antisaccade task (upper panels of Fig. 4.20)also contained more than 50% express saccades. The error rate is 35% at the left and 42% at the right side. Of these errors 87% and 92% were corrected after very short correction times of 131 ms and 129 ms, respectively (lower panels of Fig. 4.20). This indicates that the subjects have no problem of looking to the opposite side. They do reach the destination, but they get there

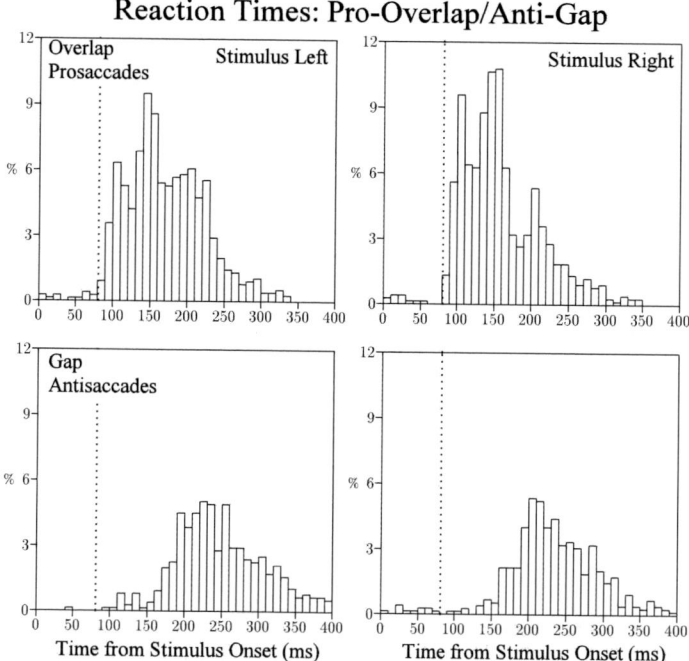

Figure 4.19. The figure shows the distributions of reaction times of 8 subjects performing the prosaccade task with overlap conditions (upper panels) and the antisaccade task with gap conditions. Panels at the left and right show the data for left and right stimulation, respectively.

with a detour because they could not suppress the saccade to the target. Their errors were mostly due to a weakness of the fixation system.

This reminds us, that we already have 2 independent factors of instability of fixation: the intrusive saccades, and the binocular instability of slow movements of the two eyes in different directions or with different velocities. Now a third aspect is added by the occurrence of express saccades and in particular, when they occur as errors in the antisaccade task. Fixation may also by weak, when it does not allow the suppression of the errors in the antisaccade task.

4.3.5. The Age Curves of Saccade Control

After these considerations and definitions we can look at the age development of the different variables. The data presented here contain many more subjects than in an earlier study, which has shown already the development of saccade control with age increasing from 7 to 70 years [Fischer et al. 1997].

Fig. 4.21 begins with the age curves of the performance of prosaccades with overlap conditions. The reaction times start with about 240 ms at the age of 7 to 8 years. During the next 10 years the reaction times become shorter by about 50 or 60 ms. From the age of

Errors and Corrections

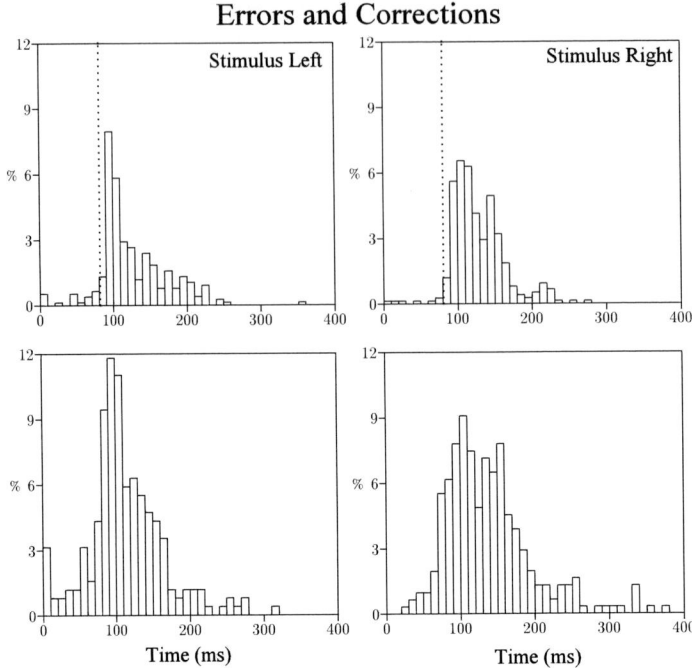

Figure 4.20. The figure shows the reaction times of the errors (upper panels) and the distributions of the correction times (lower panels) of the same subjects as in Fig. 4.19.

40 years one sees a gradual increase of the reaction times. At about 60 years they reach the level of the 7 year old children.

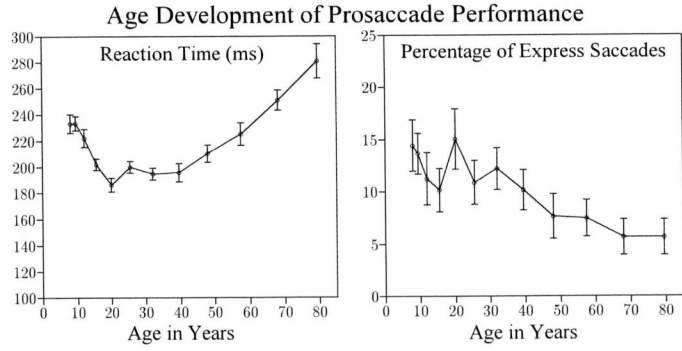

Figure 4.21. The diagrams show the age curves of the performance of prosaccades with overlap conditions. The left side depicts the age dependence of the reaction times, the right side shows the age dependence of the percentage of express saccades in the distributions. N=425.

One might expect that the occurrence of reflex-like movements (express saccades) is also a function of age, because the reflexes receive more cortical control with increasing age. However, this general aspect of the development may be seen much earlier in life, i.e. during the first year of life. Yet, there is strong tendency of a reduction of the number of express saccades with increasing age from a mean value just below 15% to a mean value of about 5%. There are however, extreme cases of subjects producing quite a lot of express saccades. The large scatter in the data is due to these subjects.

It has been stated, that percentages of express saccades above a limit of 30% must be regarded as an exceptional weakness of the fixation system. The corresponding subjects are called express saccade makers [Biscaldi et al. 1996]. An extreme case of an express saccade maker is shown in Fig. 4.22. In this subject the express saccades occur only to the right side.

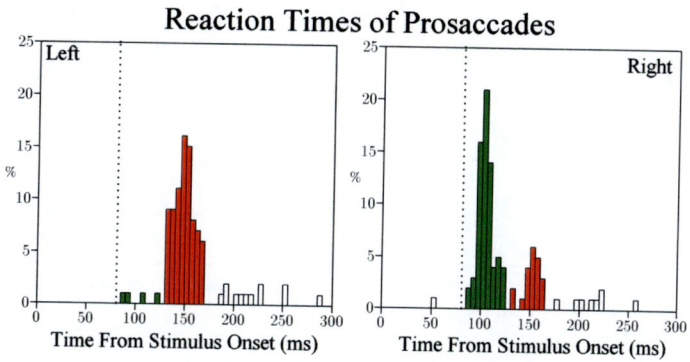

Figure 4.22. The figure shows the distributions of saccadic reaction times from a single subject, who performed the prosaccade task with overlap condition. Saccades to the left side are depicted by the left panel, those to the right side by the right panel. Note the large peak of express saccade to the right as compared with no express saccades to the left.

Later in the book we will look at the percentage of express saccades among the prosaccades generated under overlap conditions, because we want to be prepared for the diagnosis of saccade control in the following parts of the book, when large amounts of express saccades are made by single subjects of certain ages.

The Fig. 4.23 shows the age curves for the variables that describe the performance of the antisaccade task. The reaction times of the correct antisaccades are depicted by the upper left panel. The mean value of the youngest group at about 340 ms is 100 ms is slower than that of their prosaccades. As in the case of the prosaccades, a reduction of the reaction times is obtained within the next 10 years. However, they are reduced by about 100 ms. When compared with the prosaccades this reduction is two times as big.

The percentage of errors (middle left panel) reaches almost 80% for the youngest group. This means that they are almost completely unable to do the task in one step. The error rate decreases down to about 20%, stays at this level and increases after the age of about 40 years.

The bottom left panel depicts the correction rate. Out of the 80% errors, the youngest group was able to correct the primary error in only 40% of cases. The correction rate

Age Development of Antisaccade Performance

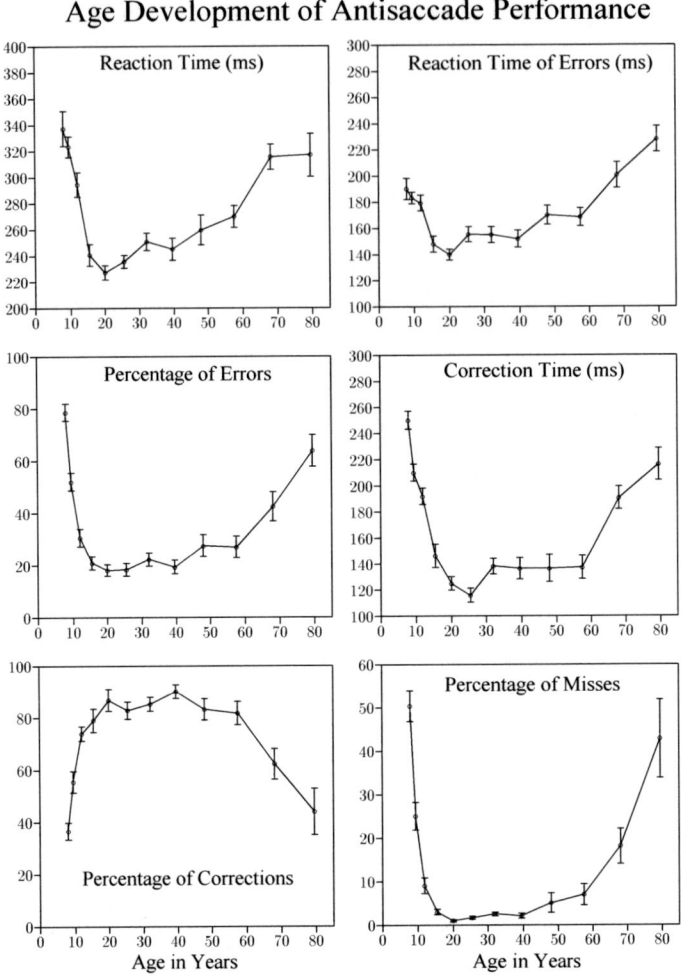

Figure 4.23. The figure shows the age development of the performance of the antisaccade task with gap conditions. N=328. The period of the "best" values is between 20 and 40 years of age.

increases until the age of 20 to above 80%, stays at this level and decreases again after the age of 50 years.

Combining the two measures of error production and correction results in the age curve of the percentage of uncorrected errors (misses) shown by the lower right panel of 4.23. The children of the youngest group reached the opposite side in only half of the trials. During the following 10 years the rate of misses drops down to almost zero. This indicates that the adult subjects in the age range between 20 and 40 years produce 20% errors, but they correct almost all of them. During years after the age of 60 the subjects begin to have more difficulties in correcting their increasing rate of errors.

Finally, we look at the reaction times of the errors shown by the upper right panel. The age curve reflects the curve for the reaction time of the prosaccades generated in the overlap condition. However, all error reaction times were shorter by about the same amount of 50 ms over the complete range of ages covered.

4.3.6.   Left – Right Asymmetries

The question of hemispheric specialisation is asked for almost any aspect of brain functions. In the case of saccade control it might be argued, that depending on the culture writing goes from left to right, right to left, or from top to bottom. Therefore we look at the possible asymmetries of the different variables describing saccade control.

The differences between left and right variables did not show any systematic age dependence, presumably because the age dependence for the right and the left variables have the same development. Therefore we look at the total distribution of the difference values for all ages.

The Fig. 4.24 shows these distributions of differences for 6 variables. The upper left panels depicts the differences in the reaction time of the prosaccades with overlap conditions. The distribution looks rather symmetrical and in fact the deviation of the mean value is only 6 ms and not significantly different from zero. However, this does not indicate that there are no asymmetries. It shows, that asymmetries occur about as often in favour of the right side as they occur in favour of the left side. The standard deviation of 30 ms to either side indicates that in 32% of the cases the reaction times differ by 30 ms or more. The tendency is that reaction times are somewhat shorter for the right directed saccades as compared with the left directed saccades. This small difference maybe related to the fact that the German language is written from left to right (all data in this book comes from native German speakers).

The upper right panel depicts the differences between the percentages of express saccades made to the right and to the left. The mean value is -1.1% and not significantly different from zero. But there is a tendency to more express saccades to right than to the left. The standard deviation if 12% indicating that 32% of the subjects produced more then 12% of their express saccades to one side than to the other. Extreme cases can be seen within this relatively large group of normal subjects. An example can be seen in Fig. 4.22. The distribution of saccadic reaction times obtained with overlap conditions are shown for left and right directed prosaccades. Almost all saccades to the left occur between 130 ms and 170 ms. These are fast regular saccades. Most saccades to the right occur between 85 ms and 140 ms. These are express saccades. The figure demonstrates an extreme case of asymmetry of prosaccades. The reaction times of the correct antisaccades in the gap condition shows a similar result: one encounters quite a number of subjects with heavy asymmetries (32% with differences of more than 45 ms), but the mean value of 5 ms is statistically not significant from zero. The correction times exhibit even stronger asymmetries: in 32% of the subject the differences are larger than 55 ms. The percentage of errors in the antisaccade task exhibit differences of more than 15% in 32% of the cases and the differences of the percentage of corrective saccades are larger than 28% in 32% of the subjects. From the consideration of the asymmetries in saccade control we can conclude that large asymmetries occur in quite many cases. Because the asymmetries in favour of the right

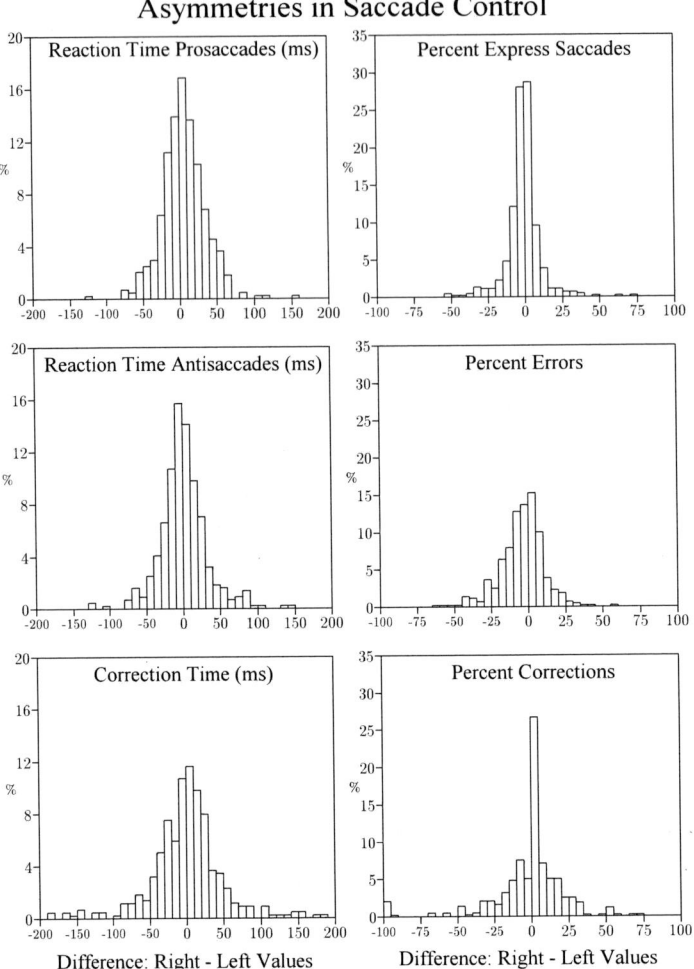

Figure 4.24. The figure shows the distributions of the left minus right differences of 6 variables describing saccade control.

or of the left side are about the same in number as well as in size, the mean value of the distribution does not deviate significantly from zero.

4.3.7. Correlations and Independence

Large numbers of errors in the antisaccade task are often interpreted as a consequence of a weak fixation system. This would imply that many intrusive saccades should be observed in the overlap prosaccade task (poor mono fixation stability) along with many errors in the gap antisaccade task. We can look at the possible correlation between these two measures. Fig. 4.25 shows the scatter plot of the data obtained from control subjects in the age range of 7 to 13 years. While the correlation coefficient indicates a positive significant correlation, the plot shows in detail, that the relation works only in one direction: High values of intrusive saccades occur along with high values of errors, but not vice versa: high values of errors may occur along with low or with high numbers of intrusive saccades. In other words: even if a subject is able to suppress intrusive saccades while fixating on a small spot, he/she may not be able to suppress reflexive saccades to a suddenly presented stimulus. But a subject, who is able to suppress the reflexive saccades, is also able to suppress intrusive saccades.

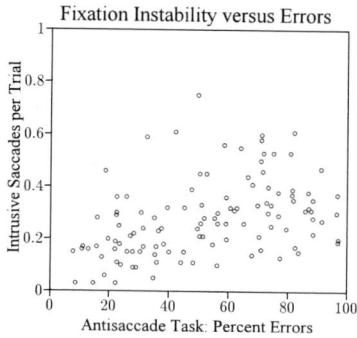

Figure 4.25. Scatterplot of error rate in the gap antisaccade task and mono fixation instability. High values of intrusive saccades occur along with high values of errors, but not vice versa: high values of errors may occur along with low or with high numbers of instrusive saccades.

This means that the reason for many errors in the antisaccade task may be a weak fixation system, but other reasons also exist such that high errors rates may be produced even though the mono fixation stability was high.

The analysis of the relationship between errors, error correction, correction time and express saccades can also be used to learn more about fixation and its role in saccade control. Those, who produce many errors and many express saccades, correct their errors more often and after shorter correction times in comparison to subjects, who produce also many errors but relatively few express saccades. They correct the errors not as often and the correction times are longer. The details are described in the literature [Mokler and Fischer, 1999].

In conclusion from this section we can state, that saccade control has indeed 3 main components: fixation (being weak or strong as indicated by express saccades), reflexive control, and voluntary control. These 3 components work together in the functional from of the optomotor cycle. The functioning of the cycle improves over the years from the age of 7 to adulthood and has a strong tendency to deteriorate after the age of 40 years [Fischer et al. 1997].

PART II

DEFICITS

Summary

This part considers the possible developmental deficits of children having specific problems at school: dyslexia, dyscalculia, attention deficits, and general learning problems. Using the variables defined in the previous part, the developmental age curves will be compared with those of control groups to see, whether significant group differences can be found. We will see, that systematic differences were found in all groups. The age curves of the mean values and their scatter (standard error) are shown. The percent numbers of children not reaching the range of the age matched control groups (defined by percentile 16, p16) are calculated.

   Sensory and optomotor brain functions are used in many ways in everyday life, minute by minute and second by second. Most of these functions have been essential during millions of years for our biological survival and many still are essential for our survival to day. The maturation of our brain together with learning processes lead to almost perfect auditory and visual perception accompanied by an almost perfect sensory motor control in general. Specifically, our optomotor control serving in vision is optimized. The fact that the corresponding neural systems function only "almost" perfect (not completely perfect) means that every once in a while the result of the processing in our brain is not quite correct. For example, we may judge the size or the weight of an object as too large or to heavy. As a consequence our motor action related to this object may be partly but not completely wrong.

   But in every day life this does not create any major problems, unless the error rate is too high or the processing systematically takes too long. For example, we may misunderstand a few words in a conversation, yet we understand, what the issue is and we can participate in the discussion without problems. Problems may occur, however, if we are trying to learn certain skills, which rely on a perfect and sufficiently fast sensory processing. Among such skills are reading, writing, spelling, or basic mathematical operations. Another example is playing an instrument: we have to read the music and activate the muscles of the fingers. In the beginning, we can easily observe the low speed of the processing in our brain. With practice we can learn to read and play the music faster. If, for example, the reading does not become faster, but the playing does, the best strategy is to learn the music by heart.

Similar to the generation of saccades auditory and visual processing may escape our subjective judgement. Therefore, it is important to be able to examine the perceptual and optomotor functions during the early years of life. It is especially important to examine them specifically during the beginning of school age in cases where one or the other learning problem occurs. In principal one can use the tasks described in the previous part for examination of single subjects to find whether or not one or the other brain function is developed within the limits of an age matched control group. One can even decide to do the examination irrespective of the kind of problem in school. If there are specific relations between the school problems and the deficits in processing, one would be able to find these relations afterwards.

However, for a systematic scientific study one wants to specify the kind of learning deficits a child suffers right from the beginning. The most common and best known problems are (i) dyslexia, (ii) dyscalculia, and (iii) attention deficits. More general learning problems may be also encountered for which a classification is not possible. This group, which is defined by exclusion (not normal and not member of one of the above specified groups) needs special consideration. These children are not eligible for scientific research, because of the difficulty of defining the criteria of "inclusion" or "exclusion" in the group under investigation. These children are not eligible for attending a regular school, either. The question, of what they are eligible for seem to be an open question in our society.

We will use "deficit" quite often in this book. Since we are dealing with quantitative measurements resulting in numbers attributed to specifies variables, we are able to calculate mean values and their scatter. A subject exhibits a deficit, when the numeric value of a variable falls outside the normal range of the control values. In general this range is defined by plus / minus one standard deviation from the mean value. This measure is equivalent to the 16 / 84 percentile of the control values, given the distribution of the values is normal. We will explicitly use the 16 percentiles as a measure of a deficit in cases, when the control data are far from forming a normal distribution as in the case of the auditory data. In the other cases we will use one standard deviation as a criterion (see Appendix).

Chapter 5

Overview and Gender

Summary

This chapter gives an overview of the different auditory and visual deficits by averaging the data across all subjects regardless of what kind of problem at school they have. The only subjects not included in this large group are the control subjects having no problems in school. Their data will be used to see whether the whole large group differs from the controls. We also look briefly at possible differences due to gender.

First, we have to find the variables allowing such a global description of the performance of the auditory, visual, and saccade tasks. From the previous part we decide: In the auditory domain we select the number of auditory tasks that could not be solved better by chance. For subitizing we select the combined variable of the effective recognition speed, and for saccade control we use the percent number of trials, in which the subjects missed to reach the opposite side in the gap antisaccade task.

Now we calculate the corresponding age curves and compared them with those of the control subjects of the task. The three pairs of curves are shown in Fig. 5.1. For easier comparison the y-axes are roughly scaled to make them all cover approximately the same range and the variables are plotted negative or positive such that all increase with age development. For example, the number of unsolved tasks are decreasing with age, while the effective recognition speed is increasing with age.

It is easy to see the similarity in the development of the deficits in the different domains. The development lasts until adult age in all domains, but is slower when compared with the controls. The deficits are relatively small at the beginning of school age and have a clear tendency to worsen with increasing age. This indicates that when school begins, parts of the brains of the children enter a new phase of learning: what was good enough for an every day life of a child until the age of 5 or 6 years, now needs to be improved to meet the new requirements.

There is always a suspicion, that all kinds of skills show a gender-specific difference. A detailed analysis of the variables does not support this notion with respects to the domains of auditory, visual, and optomotor control. But the females and males make quite different

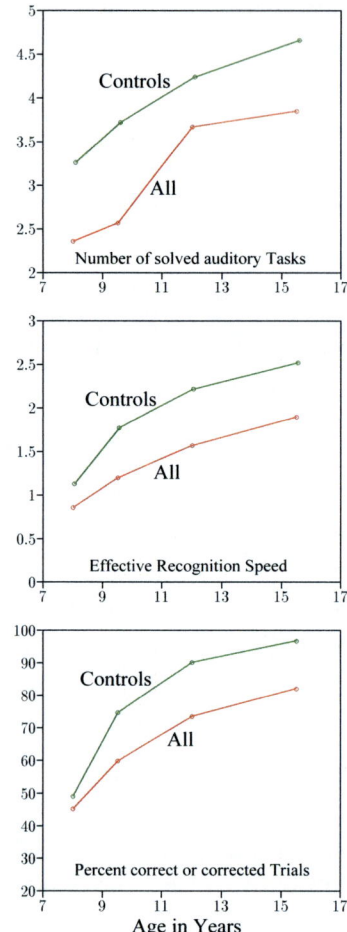

Figure 5.1. The figure shows the 3 pairs of age curves obtained by averaging the data of all subjects, who could not serve as control persons, because of one or the other deficit in learning at school. The number of subjects contributing to each domain are: N=2101 (Auditory), N=1004 (Subitizing), N=3162 (optomotor function).

contributions to the mean values of single variables to be considered below, because they appear by different percentages in the groups.

The percentage of female and male members in the different groups are presented in the Fig. 5.2. In all groups, males were over-represented as compared with females. The smallest difference is seen among the children with dyscalculia, the largest difference is obtained among the children with attention deficits. This fact, however, cannot be used to argue in favour of genetic reasons for the deficit, like dyslexia. Even though genetic factors do exist [Grigorenko et al. 1997], one would have to show, that girls and boys are differently affected. This could be assumed if the detected dyslexia-specific genes were on the X chromosome, but this is not the case.

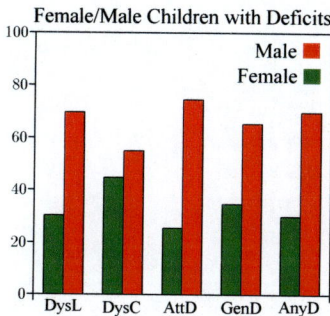

Figure 5.2. Each pair of columns shows the frequency of females among the groups shows below. DysL = Dyslexia (N=3091); DysC = Dyscalculia (N=228); AttD = Attention Deficit (N=804); GenD = General Learning Deficits (N=49); AnyD = Any Deficit (this includes all kinds of deficits that were encountered).

The different incidence rates of girls and boys among the groups and the especially large difference in the group with attention deficits are well known. We are just documenting them here to make sure that in this respect the present findings are in accordance with what is generally accepted.

Chapter 6

Dyslexia

Summary

This chapter is attributed to children with dyslexia. We will consider perceptual and op-tomotor functions that have been described in the first part. The data of large numbers of dyslexic subjects are available (N=2900) and are used to estimate the percentages among them, being impaired in one or the other function by failing to reach the limits of the control range.

6.1.   What is Dyslexia?

Since the first report of dyslexia in 1889 the definition of dyslexia remained somewhat unclear. Even the World Health Organisation (WHO) gave a description of what dyslexia is, but this could not really be used as a definition. A confusion was also created by mixing the definition with the possible (unknown) causal factors that lead to dyslexia. What is needed is a prescription of the methods that could be used in a diagnostic procedure. Only recently it became clear that dyslexia is a neurobiological deficit in learning to read and write, while the general intelligence is normal or above normal. It is also important that the child had sufficient instructions and training at school.

Today one can use tests for spelling and reading as well as an intelligence test to find the percentile reached by a single child of a certain age. A discrepancy between read-ing/spelling on the one hand and general (non-verbal) intelligence on the other is used to identify a child as dyslexic.

Of course, a child with obvious difficulties in school suffers from the deficit and even more so, when other pupils, teachers, and parents are not aware that the problem has nothing to do with intelligence or laziness. The child gives the best and yet it fails to reach the read-ing/spelling level of the class. This experience creates increasing psychological problems up to a complete refusal to learn anything at all. After a short time such a child is behind the normal development in many other aspects of school and there is a chance of confounding the causes of the behaviour of the child: low levels of reading and spelling competences are regarded as consequences of missing cooperation, low general accomplishments in all respects, or other general educational problems.

One of the advantages of having standard methods to examine the perceptual capacities using objective measures is the possibility to learn that neither laziness nor poor lack of intelligence is responsible for the learning deficit. First, the diagnosis helps to relieve the child (the parents and the teachers) from any kind of wrongly assumed liability. Second, the diagnosis opens the possibility of specific help as we will see in the part Training of the book.

6.2. Low Level Auditory Functions in Dyslexia

Children with deficits in reading and/or spelling, who did well in all other aspects in school and who reached normal or even above normal scores in an intelligence tests, were classified as dyslexic. An exact and complete diagnosis of dyslexia was not available for all the children. However, for practical purposes such exact quantitative diagnostic data are not as important, because the interest is in deficits in perception of children with reading and/or spelling problems irrespective of their intelligence.

Over the years large numbers of dyslexic children were examined for their auditory functions in the Freiburg Optomotor Laboratory. They all took the 5 diagnostic auditory tasks of volume, frequency discrimination, gap detection, time order and side order. By comparing their results with those of the normal subjects of the age matched control group, each child was assigned a percentile value (pr). The percentile is a number between 1 and 100. This number indicates at which rank order position the child appears in a rank order of the control group, when the best position is 100 and the worst position is zero. p16 corresponds to 1 standard deviation if the distribution is gaussian. p1 means, that the subject performed below the lowest level among the control subjects or even failed to solve the task at all, i.e. not better than by guessing. p90 means that only 10 percent of the control group performed better. The advantage of using the percentile is to be largely independent of the exact form of the distribution of the control values.

In a first analysis of the data we count the number of tasks (out of the 5 tasks) that were performed below certain criterion marks (limits). Each subject was assigned a number between n=0 (no tasks below, all tasks above the criterion) and n=5 (all 5 tasks below criterion). As criterion we used p16 (the classic criterion), p5 and p1 to see how severely subjects are affected.

The Fig. 6.1 shows the distributions of the variable n for the 3 criterions.

The percent numbers above the columns indicate, how many subjects performed below the criterion level, when the criterion number of tasks is increased. For example, about 23% of the subjects performed all 5 tasks above the criterion (green column). This means, that 77% of the subjects failed in 1 or more tasks the criterion of p16. About 60% of the subjects failed in 2 tasks or more. Using the stronger limits of p5 and p1 the distribution is moved to smaller numbers: more tasks are performed within the stronger limits. At least 1 task was failed in 65% and 58% of the cases, respectively. The chances that all 5 tasks failed the criterion is very low. It decreases from about 5% (p16), to 1.7% (p5), and to 0.4% for the limit of p1.

This indicates that the tasks are well understood and the poor performance is not due to an inability of performing the two-alternative-forced-choice task. The bottom panel shows, that the decrease of the limit to p1 still leaves more than half of the subjects, who were

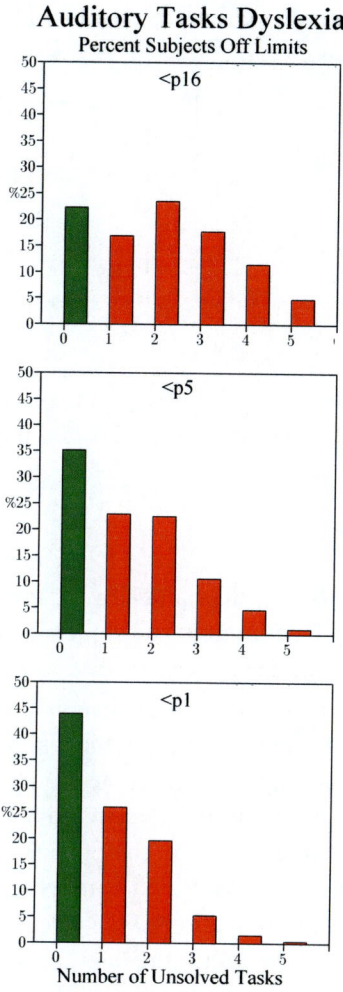

Figure 6.1. The 3 panels show the distribution of the number of tasks performed below the 3 criterions p16, p5, and p1, from top to bottom.

unable to perform at least 1 task better than p1. This fact indicates, that if a subject is impaired at all on one task (p16), the impairment is usually a heavy one.

As a next step in the analysis we look at the age development of the number of tasks that were performed below the criterion limit of p16 and p1. The Fig. 6.2 shows the results: the left side depicts the number of tasks performed off-limits for p16 and p1. The right side shows the corresponding percent number of subjects.

For p16 the number of off-limit tasks varies between 2 and 3. A clear dependence on age cannot be seen. The mean value across age is 2.5. For p5 mean value across age of 2.0. It is important to note, that even the strongest criterion of a limit p1 leaves almost 2 tasks (mean value = 1.7) below the limit.

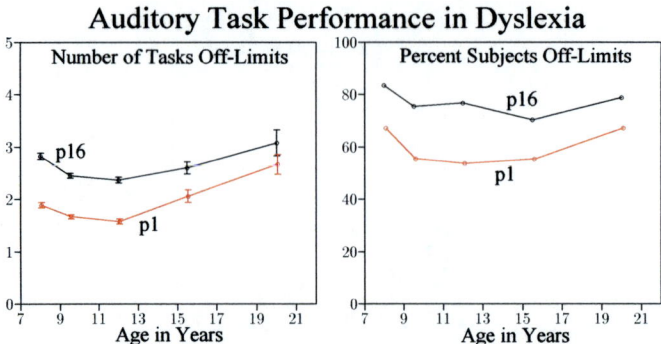

Figure 6.2. Left side: The mean number of off-limit tasks are shown as a function of age for a limit of p16 and a limit of p1. The right side shows the corresponding percent number of subjects as a function of age.

To see which of the tasks contribute how much to the low performance, we repeated the analysis separately for each task. The number of subjects below criterion were counted and averaged for each age group and for each task.

The Fig. 6.3 shows the age curves. The left panels show the percentage of subjects failing the criterion of p16. The middle and right panels shows the percentages of subjects failing criterion p5 and p1, respectively. Note, that the gradual change of the criterion (from left to right) decreases the percentage in the frequency discrimination task and in the time order tasks to a lesser extend as compared with the other 3 tasks. In fact, the tasks 2 and 4 are the most difficult tasks among the 5 described here. Studies in the literature report also that it is the fast processing part of the auditory subsystem, which creates the problem in dyslexics [Stein and Walsh, 1997]. The present results support this view combined with the extra message, that not all the dyslexics suffer from this deficit in the temporal domain of auditory processing.

Only after this analysis we can really understand the differences in the 5 pairs of age curves of the mean threshold values shown in Fig. 6.4. The average threshold values were calculated as a function of age using only those tasks, in which the subject performed better than by chance.

The differences are only marginal for some age groups in the volume-task, the gap-task, and the side order-task. The volume task yields about the same mean values for the dyslexics and the controls. Only at higher ages there is a difference. For the frequency discrimination and the time order task, on the other hand, large differences were obtained in all age groups. These are the auditory domains, in which most of the dyslexic children are most severely impaired. Deficits in frequency discrimination have also been made responsible for specific language impairments [Stein et al. 1986].

Because both of these two tasks require frequency discrimination, one might argue, that they challenge the auditory system in the same domain. However, the correlation between the individual threshold values reached a correlation coefficient of only 0.32 for the group of N=416 children, 9 years old. This low correlation is statistically significant with a p-value below 0.000, but just because of the large number of values. For practical purposes –

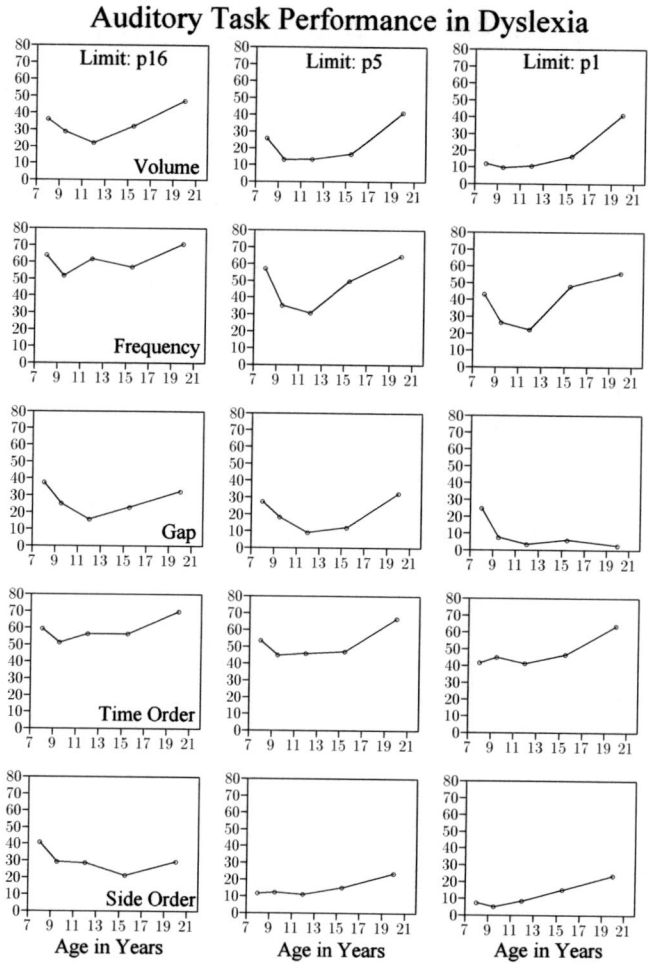

Figure 6.3. The panels shows the percentage of subjects failing the 5 auditory tasks by a criterion of p16, 5, or 1 as a function of age.

to predict one value from the knowledge of the other – this correlation does not help at all (low correlation coefficient) and we have to consider the frequency domain as independent from the time order domain. Even if one tries to predict a pr-value above 16 (below 16) reached in one domain from the pr-value reached in the other domain, one fails in about 30% of the cases (false positive plus false negative cases). There are children who reach only p0 in one task but p50 in the other and vice versa. The complete correlation matrix contains regression coefficients all below 0.32. (When calculating the correlations one can use only subjects of the same age, because all variables change with age.)

We want to close this section by trying to find an answer to the question: are dyslexic children impaired on low level auditory functions? We have to decide the criterion to be used. We have seen, that out of the tasks that were performed below a pr-limit of 16 or 5 or 1, frequency discrimination and time order were the most common, while the other 3 tasks

Auditory Task Performance

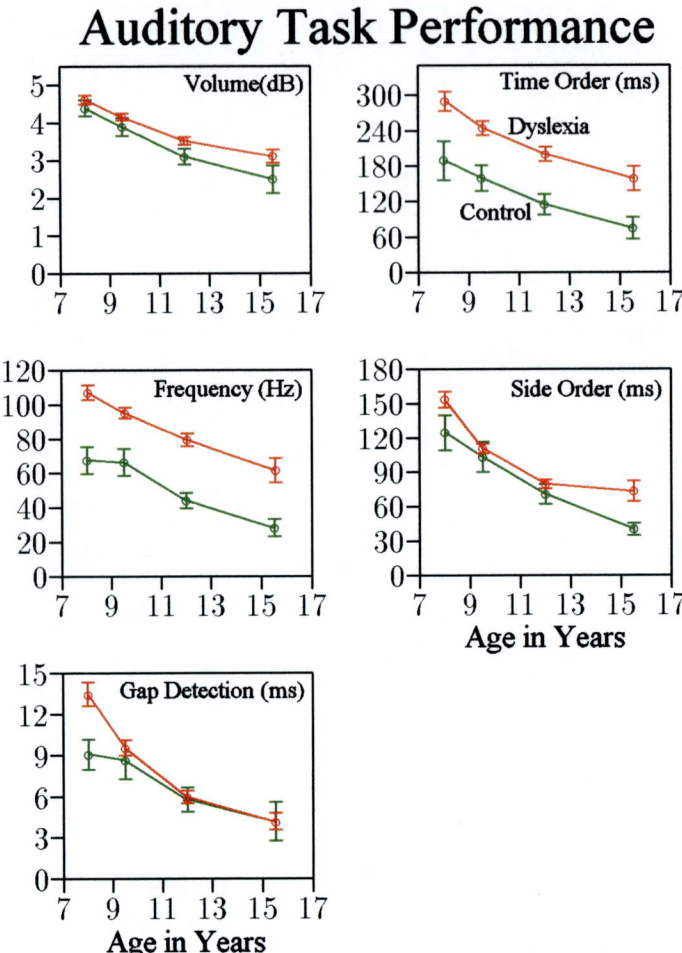

Figure 6.4. The panels show the pairs of age curves obtained from the 5 tasks. The values of the control subjects can be directly compared with those of the dyslexic children. Remember, that only those tasks contribute to the mean values, in which a subject could reach a threshold value.

were not affected as much. Therefore, we will use 2 or more tasks as a limit for the tasks and a medium criterion of p5. The result: about 65% of the dyslexic children are impaired. Whenever one wants to know the details, one has to go to the diagrams presented in this section, especially to Fig. 6.3. More details are published elsewhere [Fischer and Hartnegg, 2004].

Phonological Awareness

It has been claimed by many studies, that dyslexics suffer from auditory deficits [Mc Anally and Stein, 1996]. The previous section supports this view and shows in detail, which of the sub-domains are affected to which extend. Quite often, possible other causes for dyslexia like visual or optomotor deficits, are denied. Even within the auditory system different authors reach different conclusions.

Having seen this complexity of deficits in dyslexia in the low level auditory domain one begins to understand the possible reasons for discrepancies in the literature regarding the question, whether or not dyslexics suffer from auditory deficits. Those who did not even use the corresponding tasks, but instead used only tasks of phonological awareness come to the conclusion, that the problems of dyslexia are found in domains that analyse fundamental elements of language. This statement may be valid, as long as it just states what has been found. But it is an invalid conclusion, that the causes for dyslexia are to be localized in language processing domains, unless one has also tested low level auditory functions in the same children.

Others, who used non-linguistic tasks and did not succeed in finding significant differences between dyslexics and controls, come to the conclusion: the deficit is not in low level functions. They failed to find a deficit (in the functions that they were looking for), but this does not imply, that there is no such deficit. They just failed to find it, others did not fail, and the reasons for the failure are to be found. In a recent study both measures have been used and confirmed the result reported here.

We will see in the chapter on "Training" and "Transfer", that a group of dyslexic children, who failed the auditory tasks described here, also failed a phonological task, which used similarly sounding Germany words (rhyme words). The mean percentile of the group was 11. None of the 31 members reached percentile above 20, 23 reached p16 or less. After the training all of the 31 members reached percentiles above 20.

This suggests, that deficits in phonological awareness were caused by deficits in low-level auditory processing.

6.3.   Dynamic Vision in Dyslexia

Since dynamic vision uses the magnocellular pathway, which also provides visual information for the cortical structures that control fixation and saccades, it is of interest to examine this subfunction of the visual system in children with dyslexia. First of all, the fast sequences of saccades require a similarly fast visual process during the short periods of fixation. Second, the control of saccades relies on the same neural pathway leading from the retina to the cortical centres for saccade control.

The test of dynamic vision was applied in 511 children with dyslexia in the age range of 7 to 22 years. All 5 variables, that can be derived from the test data have been compared with those of 152 control children. The Fig. 6.5 shows the corresponding pairs of age curves.

The differences vary with age and task variable. This suggests that only relatively small percentages of the dyslexics fail to reach the p16 limit. There are considerable variations from age group to age group. The overall failure averaged across variables and age is in the order of 33%. This means that the chance of a dyslexic child to fail the test of dynamic vision is about a factor of 2 above that of the controls.

To quantify this result, we look at the mean values across the task variables. The Fig. 6.6 shows in the upper left panel the percent number of correct responses averaged across the 3

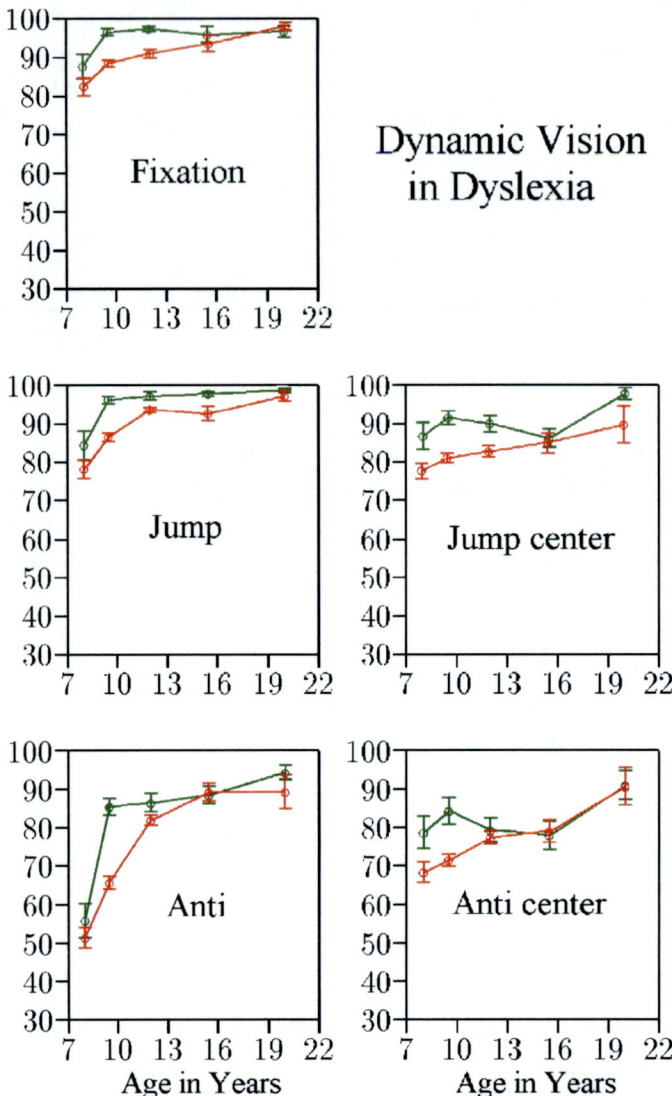

Figure 6.5. The panels show the 5 pairs of age curves obtained from dyslexic and control children.

main tasks (Fixation, Jump, Anti). The upper right panel shows the average values for the centre trials of the jump and the anti tasks. While in the age group of the 9 and 10 year old dyslexics many children failed the p16 limit, in the other groups smaller percentages are obtained. The overall failure averaged across variables and age is in the order of 33%.

Compared with the auditory domain, the test of dynamic vision differentiates the controls from the dyslexics, but the differences are not as striking. One reason is, that the test has the same difficulty for all ages. The parameters of the tests were assigned certain fixed values to make it difficult enough for the adults and easy enough for the children. This caused the subjects above the age of 15 years to experience a so called ceiling effect. This

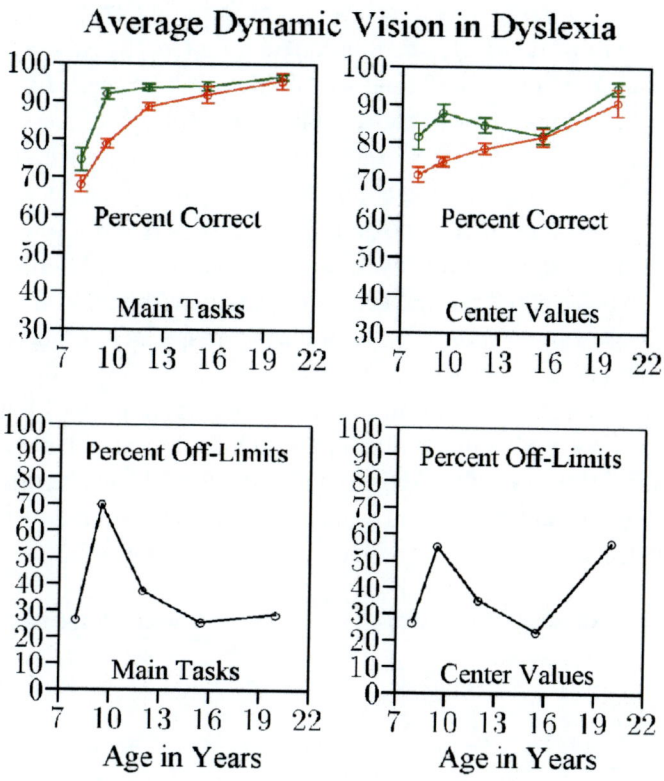

Figure 6.6. The panels show the average performance of the test tasks of dynamic vision of the controls and of the dyslexics. The upper panels depict the percentage of the correct responses, the lower panels depict the percentage of subjects failing the limit of p16.

means that for these subjects the test is so easy that a large percentage of them reaches the best possible result of the test, thus the test can not differentiate their level of performance. We have seen, that among subjects above the age of 40 years there is already a substantial percentage, who had difficulties with the tasks. This means, that at these ages ceiling effects of the tests are smaller [Fischer and Hartnegg, 2002].

Direct and indirect evidence of deficits in the mango-cellular visual system has been reported [Eden et al. 1998]; [Stein, 1993]. The present data support those reports, but also show, that not all dyslexic children suffer from deficits in the fast processing part of the visual system [Fischer et al. 2000].

6.4. Subitizing and Counting by Memory in Dyslexia

(Note: We will use "subitizing" for "subitizing and counting by memory" for the sake of brevity.)

Because subitizing deals with numbers (remember that the digit keys have to be used for responding to the presentation of the stimuli), this special visual capacity was related to arithmetic. However, subitizing may play a role in the reading process as well, because the words are combinations of relatively small numbers of letters. Most words contain 2 to 5 or 6 letters. During each fixation a small number of letters can be seen. Even in longer words, where our eyes have to make two or more stops to read the syllables (graphemes), which are then in the brain combined to one word, small numbers of letters must be identified and stored in memory. Only when thinking about the task of the visual system in reading, one may come to the question, whether or not subitizing may play a role.

The task of subitizing and counting by memory was used in children with reading and/or spelling problems, who reached normal scores or grades in the other disciplines of school. To begin with we look at the response time for correct responses for increasing numbers of items of children with an age between 9 and 10 years. The left panel of Fig. 6.7 illustrates that the response times are about the same for item numbers up to 3 but an almost linear increase can be seen for item numbers above 4. This general behaviour is the same for control children and children with dyslexia, but the curves of dyslexic children are displaced by slower response times and smaller percentages of correct responses (see right side of the figure.) This indicates already, that a developmental deficit is present for this age group.

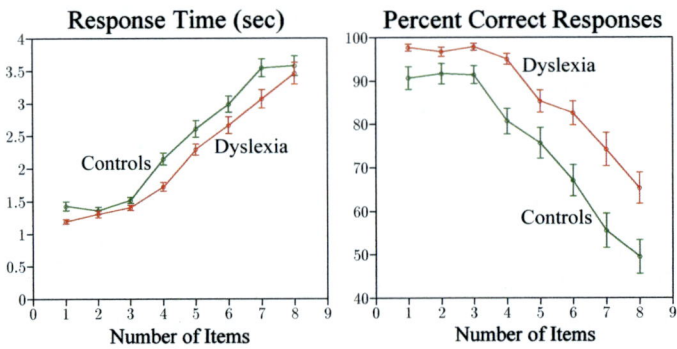

Figure 6.7. The panels show the response time and the percentage of correct responses as a function of the number of items presented. The curves are shown for 9 and 10 years old control subjects and dyslexics of the same age.

Now we proceed to analyse certain variables as to their systematic differences between dyslexics and controls.

First we analyse the variables obtained for 1 to 3 items, because we know from the previous chapter, that up to 3 items can be subitized even by young children. The Fig. 6.8 show the pairs of age curves. On the left the response times are depicted, on the right the mean percentage of correct responses. The response times of the 2 younger groups hardly differ from the control curve. Only at higher ages do clear discrepancies become evident.

The percentage of correct response is high in both groups at all ages indicating that indeed the item numbers of 1 to 3 could be subitized. But the dyslexics fail in more cases and exhibit a large interindividual scatter.

Combining the effects of both aspects of subitizing increases the difference between both groups. This can be seen when we use the derived variable of the effective recognition speed (defined as percent correct responses divided by the response time). The pair of

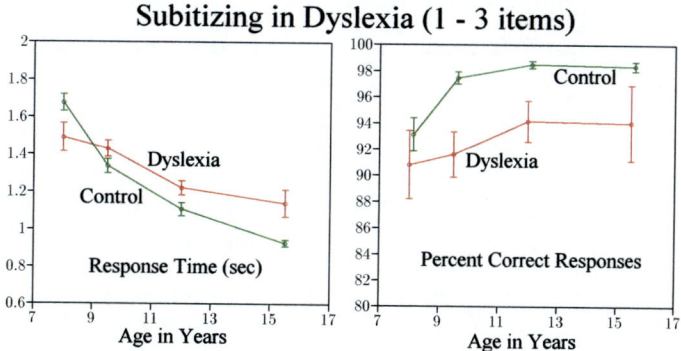

Figure 6.8. The panels show the pairs of age curves of the variables describing subitizing of 1 to 3 items. The mean response times are depicted at the left. The mean percentage of correct responses are depicted at the right.

curves is shown at the left side of the Fig. 6.9. The two youngest groups do not differ. The discrepancy increases with age. The right panel of the figure shows the percentage of subjects who failed the p16 limit. This number increases steadily with age from about 20% to almost 80%. Remember, that by definition, 16% of the controls also "fail" the p16 limit.

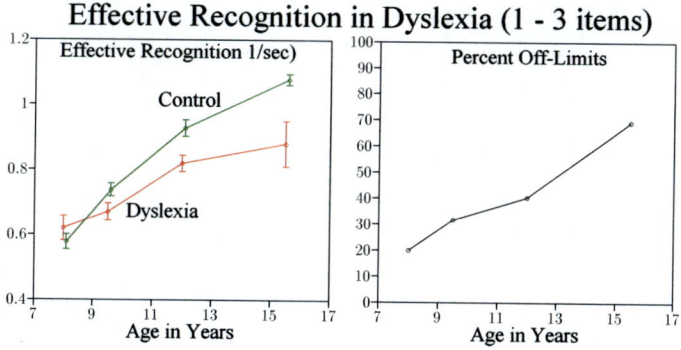

Figure 6.9. The left panel shows the effective recognition speed for 1 to 3 items obtained from the two groups of controls and dyslexics. The right panel show the percentage of dyslexic children who failed to reach the p16 limit.

The variables describing the performance of the tasks with trials of 4 to 8 items are also evaluated. The Fig. 6.10 shows on the left the time per item and on the right the mean percentage of correct responses. The time per item curve is regularly displaced to slower values and the percentage of correct responses shows smaller values at all ages.

Finally we try to estimate the percent number of dyslexic children who failed to reach p16. We decide to use the variable effective recognition speed, because it combines two variables. The Fig. 6.11 shows the effective recognition of the two groups as a function of age in the left panel. The deficit of the dyslexics can be seen at all ages. The right panel shows the percentage of subjects failing the p16 limit. The values raise with age from 30%

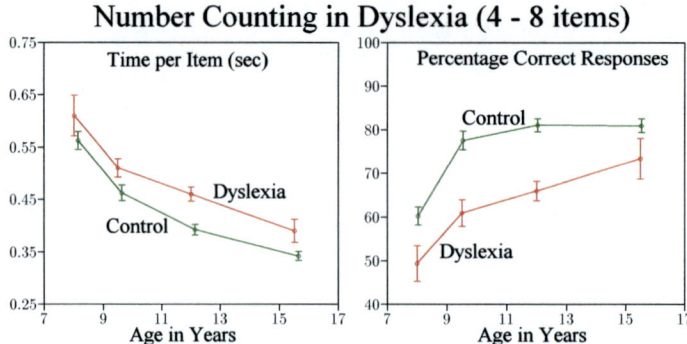

Figure 6.10. The panels show the variables obtained from the trials with 4 up to 8 items. The mean time per item is depicted on the left, the mean percentage of correct responses is depicted on the right.

to just above 70% indicating, that systematic deviations from the control group are present in the group of dyslexic subjects.

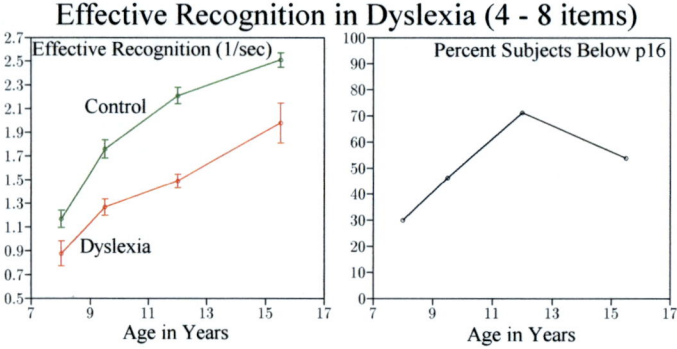

Figure 6.11. The left panel shows the age curves of the effective recognition speed for both groups. The right side shows the percentage of dyslexic children who failed to reach the p16 limit.

In conclusion to this section we have to state that the visual functions of dynamic vision and subitizing play a role in quite many dyslexic children. Even in the cases of younger subjects when the percentage of dyslexic children failing the criterion is relatively small (30%, factor of 2 as compared with the controls), one has to go through the diagnostic procedure for each individual to make sure that the child is not affected.

6.5.   Saccade Control in Dyslexia

The reading process consists of sequences of saccades and fixation periods [Rayner et al. 1983]. The size of the saccades and the duration of fixations are influenced by many factors including the physical structure of the text, the familiarity of the words in the text, the processing of the words by the corresponding centres in the brain [Rayner, 1978]. The close relationship between the reading process and eye movement control has been known for a long time [O'Regan, 1990]. The relationship between linguistic processes and saccade generation during reading has been investigated and a model of this complex coordination has been proposed [Reichle et al. 2003]. This model includes the frontal functions of saccade control and even predicts, that deficits in this special component would lead to difficulties in reading (6677). Of course, visual attention working closely together with saccade control also plays a role in dyslexia [Rayner et al. 1989].

If one examines the eye movements of dyslexics during reading, one sees irregularities of all kinds. But in this case one does not know whether these deficits arise from poor language processing (and from underlying auditory problems) or from dysfunctions of the optomotor system as such.

Therefore, saccade control was examined by tasks described in the previous part, that do not require any reading or language processing [Fischer et al. 1993]; [Biscaldi et al. 1994]; [Fischer and Weber, 1990]. Below we describe the results of the pro- and antisaccade tasks using the variables defined in the first part. Imperfections revealed by this kind of analysis of the eye movements can hardly be attributed to impairments of the language related brain processes. As we will see, certain aspects of saccade control are systematically impaired in children with dyslexia, but other aspects of saccade control are normal in most dyslexics. The answer to the question of deficits in saccade control in dyslexia, yes or no [Pavlidis, 1985]; [Pavlidis, 1985]; [Olson et al. 1983]; [Olson et al. 1991], [Rayner, 1985] is given by including all aspects of saccade control and by a thorough analysis of the data.

An example of the data from a single subject is shown in Fig. 6.12. The subject was 9 years old and suffered from dyslexia. The distributions for the reaction times of the prosaccades with overlap conditions is shown, separately for left and right directed saccades. The typical modes of express and regular saccades can been in the right part. On the left side the express mode is missing. Note also, that there are a few extremely long reaction times, which contribute to the size of the scatter. The dotted vertical line indicates the lower limit of reaction times (80 ms).

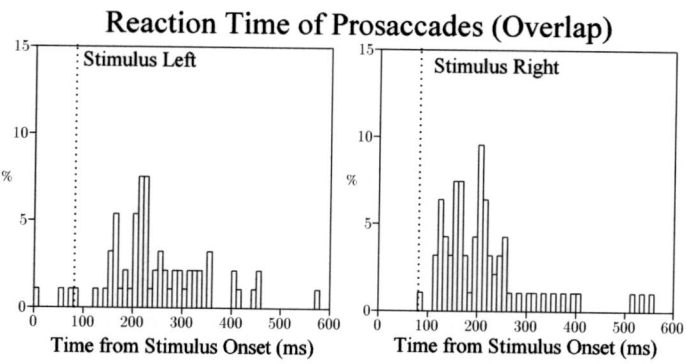

Figure 6.12. The figure shows the distributions of reaction times for prosaccades with overlap conditions, separately for left and right directed saccades, from a 9 year old dyslexic subject.

The same subject produced errors almost exclusively in the antisaccade task, 96% for both sides of stimulation. The reaction times of these errors are shown in Fig. 6.13. One sees the asymmetry again: no express saccades to the left, a clear express mode to the right. The subject corrected a small percentage of the errors (21%) made to the left side, but did not correct any of the errors made to the right side. The binocular stability was normal in this case.

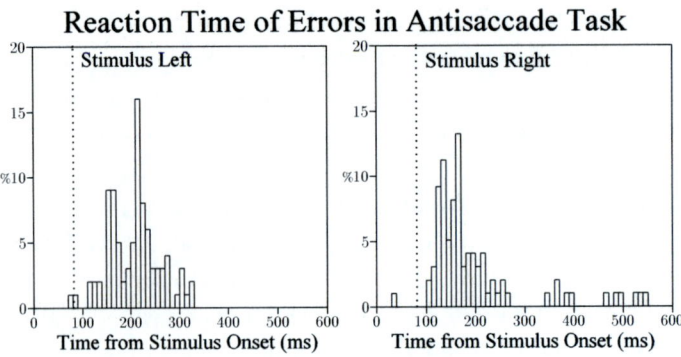

Figure 6.13. The figure shows the distribution of the reaction times of the errors made in the antisaccade task. The error rate was 96% for both sides, and the correction rate was 21% for error correction at the left and 0% for error correction at the right side.

The values derived from these distributions were used to calculate mean values and standard deviations for the members of defined age groups and show them as age curves.

The Fig. 6.14 shows on the left side the saccadic reaction times of prosaccades (overlap condition) as a function of age. The differences between the two curves are marginal. Given the scatter in the data, one has difficulties in claiming that the dyslexics exhibit systematic problems in saccade control.

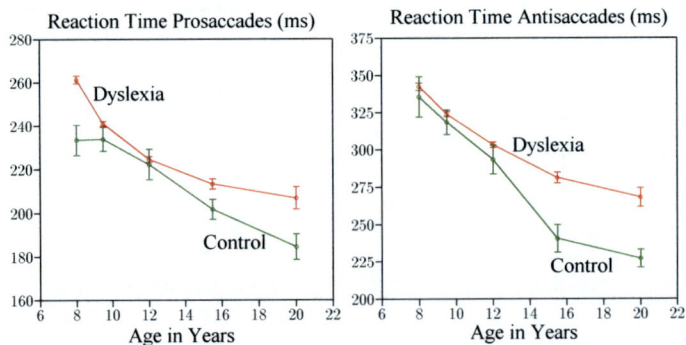

Figure 6.14. The panels show the age curves of the saccadic reaction time of prosaccades and antisaccades of the control subjects and the subjects with dyslexia.

First of all this indicates that the eye muscles are normal as well as the brain stem functions responsible for the final execution of the saccades. This view is further directly supported by analysing the maximum velocity and the size of the saccades and by comparing the almost linear relationship between these two variables (main sequence) of dyslexics and controls.

Yet, the data of a single subject may well fall outside the normal range and one wants to know exactly what the variables are and how one could possibly help the subject.

The reaction times of the correct antisaccades (right side of Fig. 6.14)are also pretty similar, at least for the younger children. With increasing age above 12 years the reaction times of the dyslexics keep decreasing but remain longer then those of the controls.

The errors made in the antisaccade task are also prosaccades (by definition) like the saccades on the left side of Fig. 6.14, that showed no difference between the groups. The Fig. 6.15 shows at the left the pairs of age curves for the pro saccades of the antitask. The younger subjects of both groups have about the same reaction times, but the older dyslexics among the subjects make their errors after longer reaction times. This variable differentiates the dyslexics from the controls even though it describes prosaccades. The dyslexics spent more time in the centre before they react and instead of using this additional time to prepare the correct reaction (an antisaccade) they still make errors. Similarly, the correction time (shown on the right side) is about the same for the younger subjects, but the older dyslexics need more time to correct the initial error. This indicates the difficulty that these dyslexics have to make a saccade to the side opposite to the stimulus.

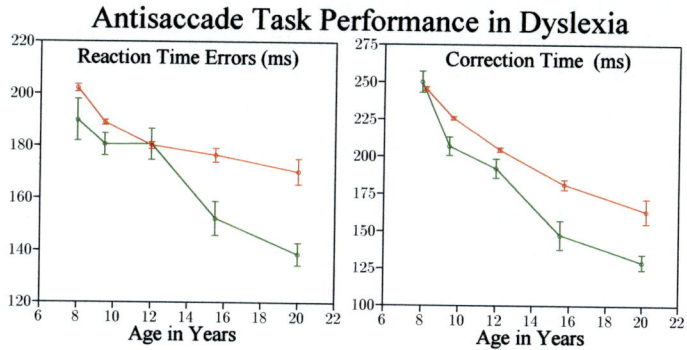

Figure 6.15. The panels show at the left side the age curves for the reaction times of the errors and at the right side the correction times for control subjects and subjects with dyslexia.

Now we look at the number of errors made. In the antisaccade task with gap conditions even control children make surprisingly many errors by looking to the stimulus first. This movement was not planned and not required. By contrast, this movement was forbidden. The question whether dyslexic children would make even more errors is answered by the Fig. 6.16. The left side shows the error rates in percent. While both groups start with high error rates (80%) at the age of 7 to 8 years, the error rate decreased quickly for the controls but only slowly for the dyslexics.

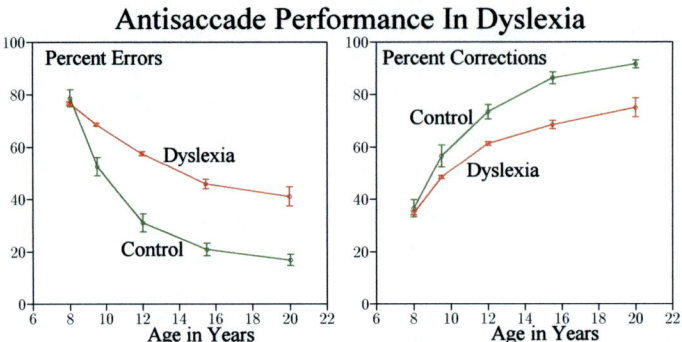

Figure 6.16. The age curves of the error rates (left) and the correction rates (right) are shown for the control group and the group of dyslexic subjects.

The percentage of corrective saccades shows a very similar development: starting with low values the correction rate increases, but it is considerably slower for the dyslexics than for the controls.

The error rate (per) and the percentage of error corrections (pcor) can be combined in a variable called misses:

$$= \ ^a\!\textit{AE}\,(100 - \ \pm\!\textit{AE}/100$$

. This variable indicates in how many trials the initial error is not corrected, i.e. the subject did not reach the opposite side, neither after one nor after two saccades (corrected error). Most of the time this means that the subject did not reach the opposite side by the end of the trial sometimes it means, that the subject reached the opposite side only after 3 saccades.

The Fig. 6.17 shows the pair of curves at the left side. Both groups start with high percentages of misses. By the variable pmis the two groups are most easily differentiated, but the values of the youngest groups are close together.

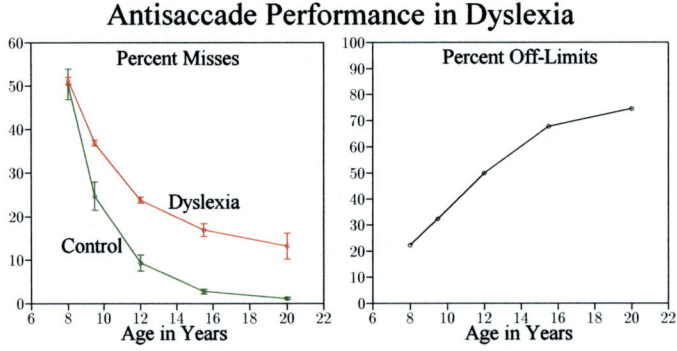

Figure 6.17. The figure shows the development of the rate of misses (left side) and the percentage of dyslexic subjects, who failed the p16 limit of the controls.

To see how many subjects among the dyslexics give rise to the differentiation we used the p16 percentile as a limit in each group. Fig. 6.17 shows that with increasing age more and more children with dyslexia could not reach the p16 limit of the controls. The highest value of more than 70% is reached by the oldest group.

When performing the same analysis for the reaction time of the correct antisaccades, smaller numbers are obtained. In particular, the younger dyslexic children are as fast as the age matched controls. Only the older groups are slower. This is also reflected in the percent number of children failing the p16 limit (data not shown here).

Therefore, we can conclude that it is the antisaccade task, which allows to distinguish the dyslexics from the controls in the eye movement domain. The prosaccade task also allows this differentiation but only for dyslexics of a certain age range. It is very important to make this distinction in the diagnostic value of the two tasks. Quite often it is said, that saccadic eye movements are the same in controls and dyslexics. Other authors say: saccadic eye movements are different. Both statements may become more correct, if one adds the task which has been used, the age of the children that were tested, and the number of subjects, that did not reach the limits of the controls group. One may firmly believe, that this simple requirement for valid scientific statements is one of the main reasons for the apparent discrepancies found in the literature regarding the question of oculomotor deficits in dyslexia. The issue of deficits in saccade control is extensively discussed in the published work [Biscaldi et al. 2000].

Of course, attention may also be a factor in dyslexia. Extra efforts are needed to distinguish attention deficits from deficits in saccade control. In this context one has to remember, that attention and the generation of saccades are closely related as has been reviewed earlier [Fischer and Breitmeyer, 1987].

6.6. Fixation Instability in Dyslexia

The disruption of fixation by unwanted (intrusive) saccades (mono-instability) has also been measured in dyslexic children and the results are compared with the controls in Fig. 6.18. While both age curves start at about the same point, the number of unwanted saccades per trial decreases steadily with age in both groups. The decline is significantly slower in the dyslexic group [Fischer and Hartnegg, 2000].

This indicates, that the line of sight of many dyslexics is not as stable as in the controls. One can find the percent number of affected subjects among the dyslexics by counting the cases that fail the p16 limit. The result is that the youngest age group of dyslexics contained no children performing outside the range of the controls. With increasing age the percentage increased from about 25% to about 40%. These numbers indicate that the problem of fixation instability due to unwanted saccades may play an important role for the dyslexics age 9 and older. During reading, however, stable fixation is needed only for short periods of time in the order of 100 to 400 ms. Yet, an intrusive saccade at an inappropriate instant of time to an inappropriate place in the text may interrupt the reading process in too many cases. If interruptions of this kind happen too often during the learning of reading the total process may suffer and during the following years the accomplishments in reading are much too low.

During fixation the angle of convergence between the two eyes should be constant, thus the relative velocity between the right and the left eye should be zero, to guarantee perfect binocular vision at the distance of the fixated object. Yet, as we have already seen, this is by

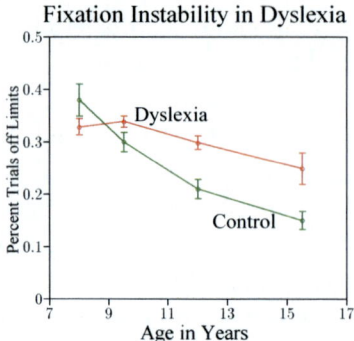

Figure 6.18. The pairs of curves show the age development of the fixation instability in control children and children with dyslexia. The data are obtained from the prosaccade task with overlap conditions.

no means always the case. Here we want to compare the binocular stability in the dyslexics and the controls. The Fig. 6.19 shows the pair of curves.

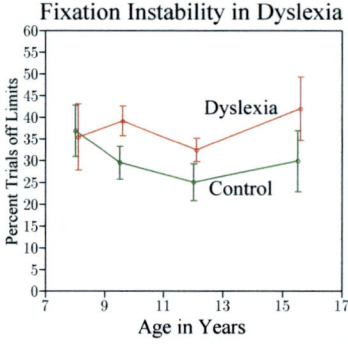

Figure 6.19. The index of binocular stability is shown as a function of age for the dyslexics and the controls. The data are obtained from the prosaccade task with overlap conditions.

Both the control values and test values exhibit large scatters: there are surprisingly many subjects, not only among the dyslexics but also among the controls, who have considerable problems maintaining both eyes directed to the fixation point. The percentage of dyslexic subjects reaching values outside the normal range can be estimated as 25 to 35%. This relatively small number is mainly due to the large scatter in the control data. Deficits of dyslexics in binocular vision as revealed by the Dunlop-Test have been reported earlier [Stein et al. 1987]. The present data support this notion insofar as binocular instabilities may disturb stereopsis. We will also see later, that monocular training (with one eye covered) reduces the binocular instability. Covering one eye during reading was also the strategy for remediation used in cases of failing the Dunlop synoptophore test [Stein and Fowler, 1985].

In a separate study the eye dominance of the binocular instability was measured: is it possible, that the dyslexic prefer one of the two eyes as the guiding eye, while the other eye produces the instability? The Fig. 6.20 shows the distribution of the differences between the right and the left values. The general rule is, that it is not possible to blame one or the other eye alone for producing the instability. Furthermore, only in a small percentage of the cases the left eye was moving while the right eye was maintaining its line of sight.

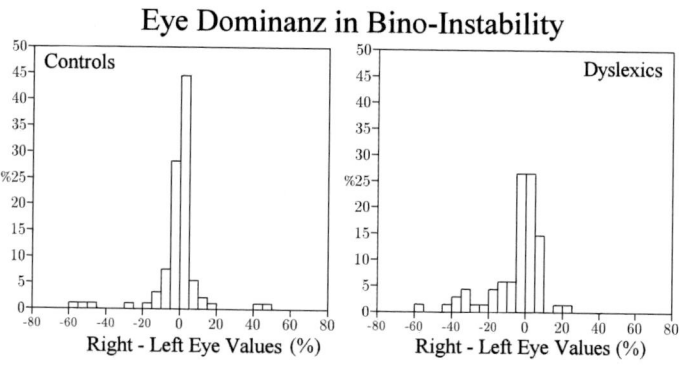

Figure 6.20. The distribution of the difference of the right and the left eye induced instability. While the mean value is negative, the standard deviation is large. There are only 8 cases where the right eye was extremely dominant in leading, the left eye producing the instability almost completely alone.

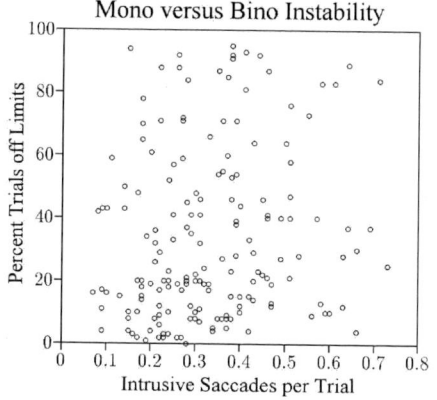

Figure 6.21. The scatter plot of the variables describing the mono and the binocular instability are shown for an age group of 9 and 10 year old dyslexic children (N=175). The correlation coefficient is r=0.13, the significance level is 0.08.

The independence of the two kinds of instability of fixation can be seen by looking at the scatter plots of the two variables in Fig. 6.21. What was true for the control subjects applies also to the dyslexics: the two kinds of fixation instability are independent of each other. When we look at the effects of eye movement training we will find more support for this independence, because monocular training will improve the binocular stability quite effectively it fails to improve the mono stability in most cases.

Chapter 7

Dyscalculia

Summary

This chapter is attributed to children with dyscalculia. The results of the examination of perceptual and optomotor functions that have been described in the first part and the comparison with the control groups will be presented. The data of subjects with dyscalculia will be used to estimate the percentages among them, who failed the p16 criterion in one or the other functional sub-domain. Different amounts of deficits are found, especially in the domain of subitizing and number counting.

7.1.   What is Dyscalculia?

Similar to dyslexia, the specific deficit in basic arithmetic skills with normal or above normal intelligence scores is called dyscalculia. Typically these children are impaired on even very simple calculations like addition of small numbers. The task of adding 3 to 4 is typically solved by counting on 3 and then – using the fingers – count further on to 7. The children know the digits – these are the visual signs of the numbers – and they know the words for the numbers –these are the auditory signals of the number. What seems to be missing is the internal presentation of how many items are related to the digits and/or number words.

We will present the data from children with dyscalculia and compare them with those of age matched controls. We have data from children in the age range from 7 to 18 years.

7.2.   Auditory Functions in Dyscalculia

As in the case of dyslexia a number of subjects with dyscalculia were unable to perform the auditory tasks better than by guessing. Others could do the task but failed to reach the p16, the p5 or the p1 limit of the controls. The Fig. 7.1 shows the distribution of the number of tasks that were performed below these limits. The first column indicates the number of subjects, who could perform all 5 tasks within the limit. This number increases, of course, as the selected limit is stronger (from top to bottom in the figure). The second and the third column show the number of subjects failing in 1 and in 2 tasks, respectively. The other columns present the number of subjects, who failed in 3 to 5 tasks. Note, that even with the strongest criterion (p1) only 35% of the subjects succeeded in all 5 tasks.

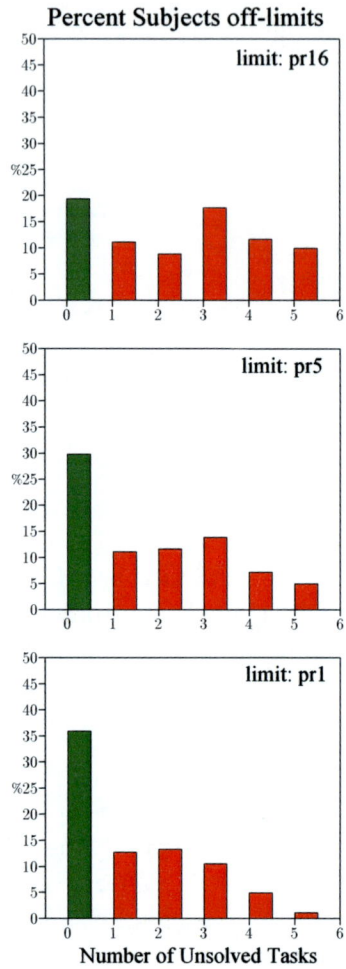

Figure 7.1. The figure shows the distribution of the number of tasks performed below the limits of p16, p5 and p1 as indicated.

The mean value of the below limit tasks is also shown as a function of age in Fig. 7.2. The age curves are shown only for limit p16 and limit p1. Age plays only a minor role for this variable. Even with the strongest criterion the mean value stays at 2 or above. The percentages of subjects are shown at the left of the figure as a function of age. In conclusion from this consideration we have to state, that auditory deficits as diagnosed by the present method are a severe problem in dyscalculia and these problems remain severe with increasing age.

The Fig. 7.3 shows the age curves of the percentage of subjects failing the 3 criterions for each of the 5 auditory tasks. The left panels show the percentage of subjects failing the criterion of p16. The middle and right panels shows the percentages of subjects failing

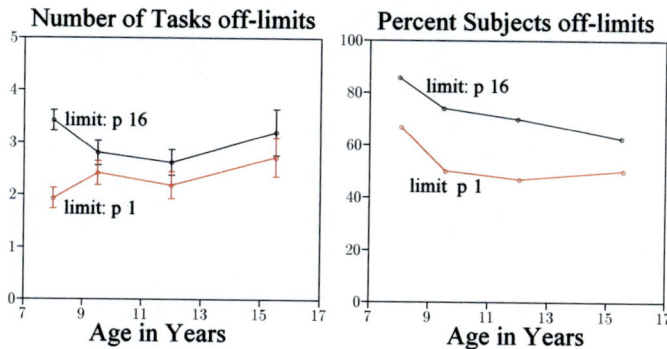

Figure 7.2. The figure shows the mean values of the unsolved tasks at the left side and the percentage of subjects who failed the limit of p16 and p1.

criterion p5 and p1, respectively. The gradual increase of the strength of the criterion (from left to right) decreases the percentage of subjects. But even with the strongest criterion of p1 a considerable number of subjects still remained as being impaired on the task. The frequency domain and the time order domain were quite strongly affected at all ages: the percentage of subjects does not fall below 30% in these two tasks.

The age curves of the threshold values are shown in Fig. 7.4. The development with age can be seen in both groups and all tasks. As in the case of dyslexia the frequency and the time order domain are most severely affected.

7.3.   Visual Functions in Dyscalculia

The visual functions described in the previous part are dynamic vision and subitizing. We will skip dynamic vision in this section, because we will treat the optomotor problems and the corresponding training of children with dyscalculia later. The training will utilize the task for dynamic vision and we will see, what can be achieved in this domain of vision anyway.

The idea that children with dyscalculia may have problems with subitizing and number counting by memory relies on the fact, that these children as a rule know the digits and the number words. Yet, when asked to add 3 to 4 they use their fingers to count from 1 on 3 using their fingers and then count further on to 7 using the fingers of the other hand. This way they arrive at the correct result, but they used a concept of the number one only. The concept, which does not really deserve this word, allows enumeration up and down in steps of one only.

If the visual capacity to see the number of items is of help for the development of the concept of number, and if a large number of children with dyscalculia did not develop this concept, one should be able to diagnose this deficit by using the task of subitizing and number counting by memory described in the first part of this book.

The Fig. 7.5 shows the response time and the percentage of correct responses for increasing numbers of items for a group of controls and a group of children with dyscalculia

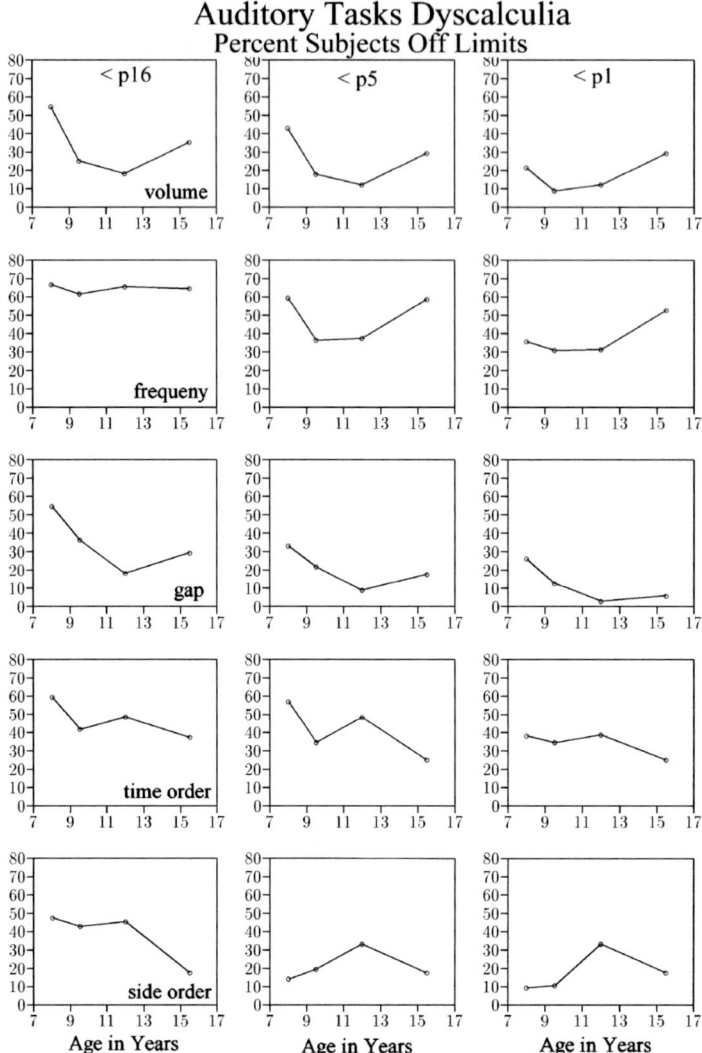

Figure 7.3. The panels shows the percentage of subjects failing the 5 auditory tasks by a criterion of p16, p5, or p1 as a function of age.

(the children were 9 and 10 years old). The response times of the test children appear to be displaced vertically to longer response times. The percentage of correct responses is also displaced to lower values, but the deficit is greater at higher numbers of items. Note that the differences between these two groups of 9 and 10 year old subjects is evident even for small item numbers, below 4. To tell the difference between one item and another number of items the children with dyscalculia need more time than the controls. Also, they make more errors: while the controls come very close to 100%, the children with dyscalculia reach only 90%. This difference might be regarded as small, but it is interesting to note, that the dyscalculics make some errors, when the controls are sure and make no errors.

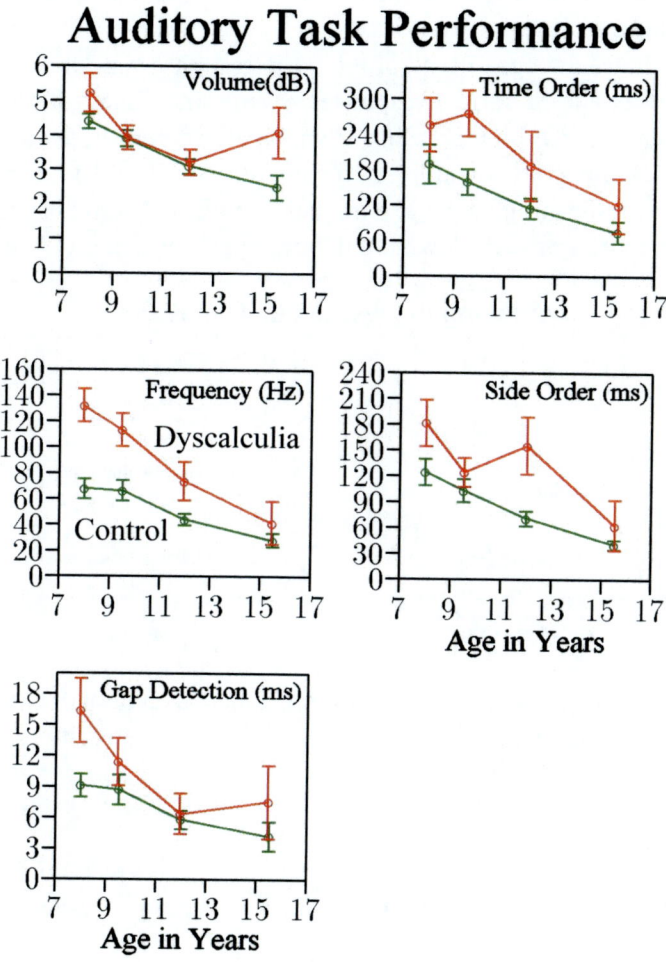

Figure 7.4. The figure shows the pairs of ages curves of the threshold valued reached by the children with dyscalculia as compared with those of the controls.

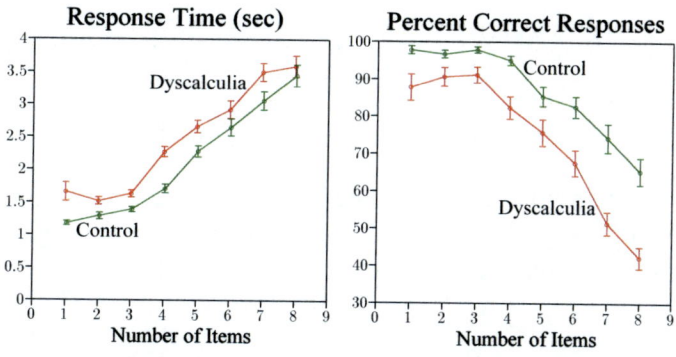

Figure 7.5. The panels shows the response time and the percentage of correct responses for a group of controls and a group of children with dyscalculia. Age: 9 and 10 years.

We will consider the different variables that can be obtained to describe the quality of the task performance.

The Fig. 7.6 shows the age curves of the basic response time for 1 to 3 items and the corresponding percentage of correct responses. While the youngest groups do not differ the groups of older ages exhibit increasing amounts of differences. Note, that the response time for one item is almost identical to the mean value for item 1 to item 3. Using t1, the number of subjects performing above p16 increased with age from 30% to 60%. The mean percentage of correct responses is lower at all ages but the interindividual scatter is large.

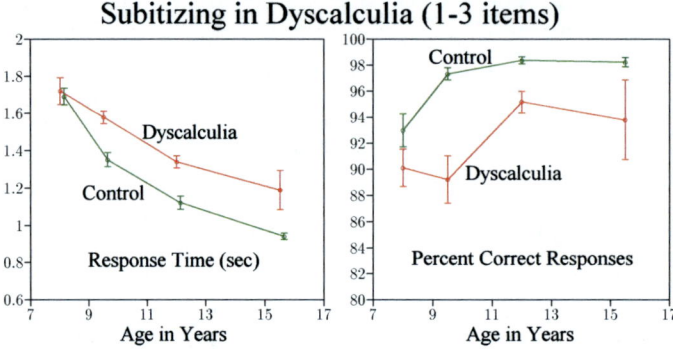

Figure 7.6. The panels show the basic response times and the percentage of correct responses for 1 to 3 items obtained from the groups of children with dyscalculia and the control subjects.

One can combine both variables by computing the effective recognition speed and count the percentage of subjects failing the criterion of p16. The Fig. 7.7 shows the results.

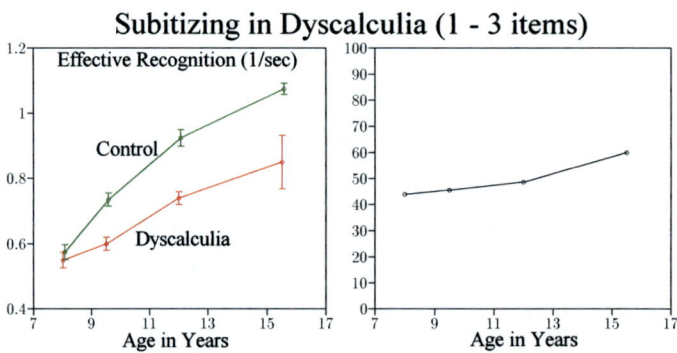

Figure 7.7. The panel at the left shows the age curves of the effective recognition speed calculated for item numbers 1 to 3. The right panel shows the percentage of children with dyscalculia failing the criterion of p16.

The two curves begin to diverge after the youngest age group. The percentage of subjects failing the p16 criterion increases slightly from 44% to about 60%. These result support the original idea, that children with dyscalculia have considerable problems with subitizing and one can imagine, that they could not use this visual capacity to associate the words and digits with the quantity of items (the number itself).

The variables describing the counting part of the task (item numbers 4 to 8) are shown in Fig. 7.8. The left panel depicts the time per item and the right panel shows the mean percentage of correct responses. Both variables differentiate the groups at all ages. The children with dyslexia need more time for each extra item and they make more errors. They could not use the extra time to produce more correct responses. The general rule of accuracy-time-trade-off does not apply in this case. The fact, that even normal subjects could not reach percentages of correct responses above 80 does not mean that they were unable to count correctly on 8. One has to remember, that the items were presented for 100 ms only.

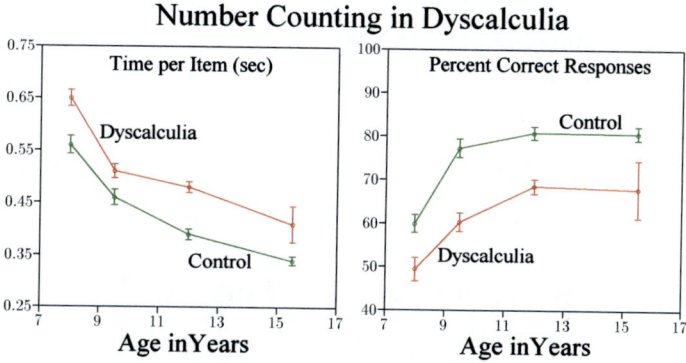

Figure 7.8. The panels show the age curves of the time per item (left side) and the mean percentage of correct responses for item number 4 to 8.

Combining the two variables we look at the effective recognition speed. The pair of curves is shown in Fig. 7.9. These age curves show the clear differences between the groups at all ages. This result is also reflected in the percentage of subjects failing the p16 criterion as can be seen at the right of the figure. Up to 80% of the children with dyscalculia failed the p16 criterion. This means that deficits in subitizing is very common in dyscalculia, but does not mean, that subitizing is the only problem they have, and it does not mean that improvements of subitizing alone will solve the problems of children with dyscalculia [Fischer, 2005]. What can be achieved will be discussed later in the corresponding chapters of the part "Training" and "Transfer".

7.4.   Saccade Control in Dyscalculia

Children with specific problems in learning basic arithmetic were examined by exactly the same optomotor tasks as described in the previous sections: prosaccade task with overlap conditions and antisaccade task with gap conditions.

The performance of both prosaccade and antisaccde task is characterized by the reaction times. The Fig. 7.10 shows the graphics at the left and the right side, respectively. While

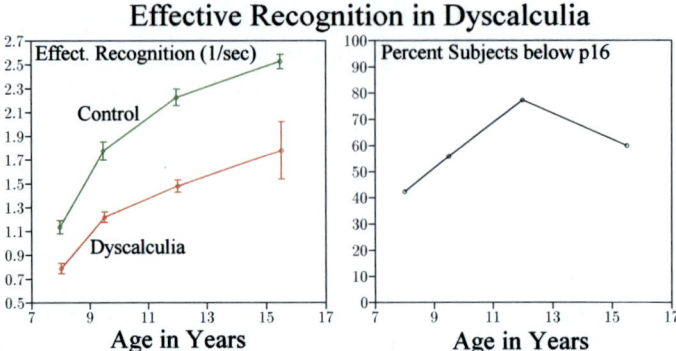

Figure 7.9. The panel on the left shows the effective recognition speed for item numbers 4 to 8 obtained for the children with dyscalculia and the controls. The right panel depicts the percentage of subjects failing the p16 criterion.

the prosaccades were made after significantly longer reaction times, the antisaccades have about the same mean reaction times with the exception of the oldest age group.

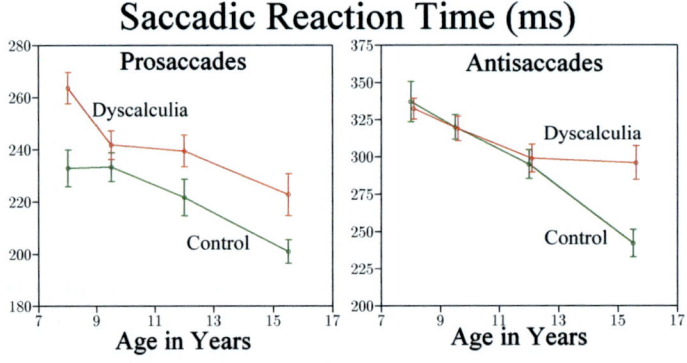

Figure 7.10. The panels shows the reaction times of the pro- and antisaccade task of children with dyscalculia in comparison with those of the controls.

The errors were made at certain reaction times and the errors were eventually (not in all cases) corrected by a second saccade after an interval, the correction time. Both variables are shown as a function of age in Fig. 7.11. The reaction and the correction times are generally slower in the children with dyscalculia. The discrepancy is largest for the oldest group in both graphs.

The percent number of errors and the percent number of corrective saccades following the errors are shown in Fig. 7.12. Like in the case of dyslexia the groups start their development at about the same level. But then the control group develops faster than the group of children with dyscalculia. Not only do they make more errors, they also correct them in fewer cases when compared with the controls of the same age.

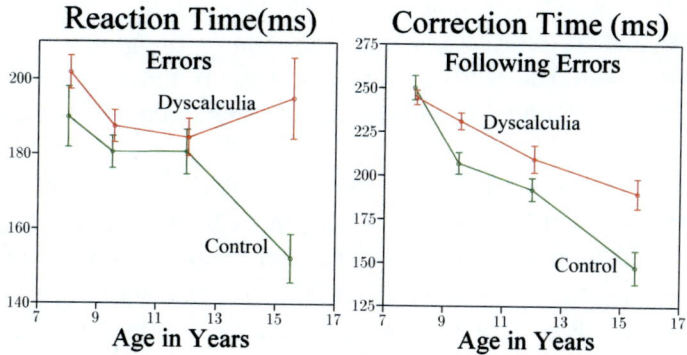

Figure 7.11. The panels show the age curves for the reaction time of the errors (left) and the correction time following the errors (right) obtained from children with dyscalculia.

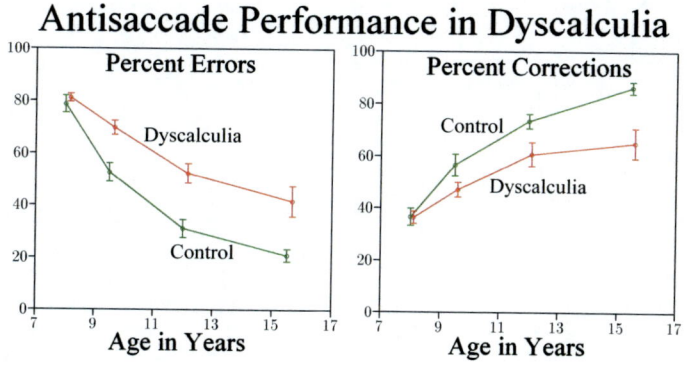

Figure 7.12. The two panels show the frequency of errors and the frequency of corrective saccades following the errors as a function of age.

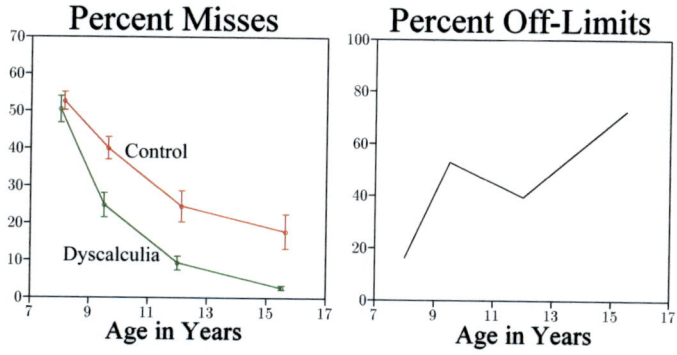

Figure 7.13. The percentage of uncorrected errors (misses), are shown as a function of age. The right panel shows the percentage of dyscalculia subjects who did not reach the p16 criterion of the age matched controls.

The combination of the error rate and the correction rate results in the frequency of uncorrected errors, called misses. The Fig. 7.13 shows the age curve of the misses obtained from children with dyscalculia. The right diagram indicates the percent number of dyscalculia subject performing the antisaccade task below the p16 criterion measured by the percentage of misses.

In conclusion, we have to state, that children with dyscalculia may also suffer from deficits in saccade control, mainly in the component controlled by the frontal brain. If these children had problems with reading the digits and/or counting small numbers by scanning saccades, one can imagine, that they may run into problems with basic arithmetic as well. But as in the case of dyslexia, we cannot expect a one-to-one relationship between the school problems and the problems with auditory, visual, and optomotor deficits. The fact, that parietal functions are involved in number processing [Dehaene et al. 2003] may point to deficits of these functions in dyscalculia. Since the parietal cortex and its projections to the frontal lobe also participate in the control of fixation and saccades, the deficits reported here do not come as a surprise.

Chapter 8

Attention Deficit

Summary

This chapter describes the data obtained from children with an attention deficit syndrome (ADS). Some also suffered from hyperactivity. The main criterion for being included in this group was the deficit in attention and/or concentration. Not all children were actually examined on their attention span or the corresponding data were not available. Excluded from this group were all children with dyslexia or with a suspicion of dyslexia. Children who used medications were required to abstain from taking it at least 6 hours before the examination and take their pill only after it. We will see, that especially the frontal component of saccade control exhibits systematic deficits in children with ADS.

8.1.   What is Attention Deficit?

The definition of the disorder is not trivial. The term "attention deficits hyperactivity disorder" (ADHD) means a complex often uncontrolled behaviour of a child, which is sometimes hard to distinguish from a child who not received an adequate education. At the behavioural level hyperactivity is evident by uncontrolled movements that the child has difficulties to stop on command. It looks as though the child is driven by all kinds of reflexive movements which are poorly coordinated and can hardly be suppressed.

This deficit also leads to the picture of a child, who cannot concentrate on the performance of tasks like doing home work for school. This fact has led to the notion, that the child's attention cannot be focused for longer periods of time. Yet, when the child decides on his own to play with Lego bricks, for example, it may be possible that concentration lasts for an hour or even longer, comparable with normal children.

It is even more difficult to identify attention deficits that are not accompanied by hyperactivity, because the child might sit quietly in the class room and seem to follow the teacher. Only when asked a question it becomes clear, that the child did not follow anything but was just sitting there, presumably with its thoughts wandering all around.

Today one knows that attention deficits and hyperactivity are problems arising from dysfunctions of the frontal or prefrontal lobe. The neural transmitter dopamine is discussed as playing an especially important role. On this basis medications have been developed to improve the frontal functions and thereby to help the coordination and control of movement.

While some of these medications work for some children, they do not work for others. Doctors and parents have taken extremely negative or extremely positive positions with regard to the usefulness of the medication, sometimes of any medication. A better way is to give it a try and use the medication in cases where it helps without too many unwanted side effects.

Here we are interested in the performance of the perceptual and optomotor tasks described above. If the deficit in attention is general then we expect that all the children with ADS will be more or less impaired in all the tasks. We will see, that the deficits do not occur in all domains. We expect that the control of antisaccades will be poor [Klein et al. 2002], because correct antisaccades need intact frontal functions [Guitton et al. 1985].

Since we do not have complete diagnoses in all cases the children contributing to this chapter are characterized by their attention deficit only (ADS). Children with ADS, who were also suffering from dyslexia, were excluded from this analysis.

8.2.   Low Level Auditory Functions in ADS

The children with a diagnosis of ADS are characterized by attention deficits: they have great difficulties in focussing their attention to a task, that is given to them, e.g. by the parents or the teacher. In the school some of these children would also lose their attention. They would just sit there quietly. In addition many of them (but not all) are also suffering from hyperactivity. Quite a number of the children would also have difficulties in reading and spelling. In this section we will describe the results of the auditory tasks

As in the case of dyslexia not all children with ADS were able to enter one or the other of the 5 tasks and perform it at its easiest level. They reached percentiles of zero. Others reached a percentile of 1. The percentage of children scoring below p1, p5 and p16 is plotted against the number of the corresponding tasks in the Fig. 8.1.

It may not come as a surprise that the percentages are as high as can be seen in the figure. Because we are dealing with attention deficits, a general attention deficit should lead to impairments irrespective of the type of task. We will see below, that this is not the case.

The Fig. 8.2shows the age dependence of the number of tasks with scores below p1 and p16 a the left side. The corresponding percentages of subjects are shown at the right side.

We also want to know, which of the tasks are the most difficult ones. The Fig. 8.3 shows all the details. While the tasks of volume discrimination, gap detection, and side order seem to cause difficulties in a smaller number of cases, the frequency discrimination and the time order tasks were failed by many more subjects. In particular, the time order task was failed by many subjects even when the strong criterion of p1 is used.

Finally, we look at the mean of the threshold values reached as a function of age for all 5 tasks. The Fig. 8.4 shows these data. Note that the question of auditory deficits in ADS does not receive a simple answer. Depending on age, on the tasks, and on the criterion used, the performance is quite different. But as a general rule one can conclude from this section, that a child with ADS should be tested for the low level auditory discrimination.

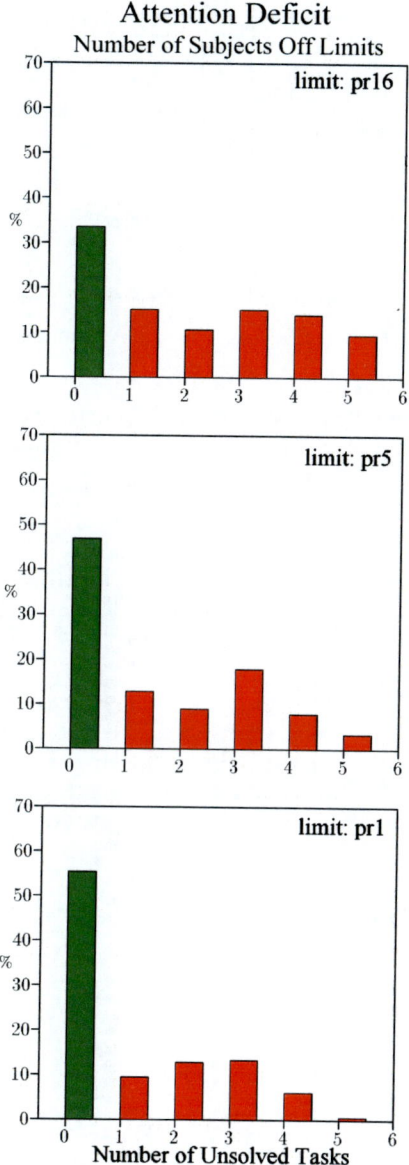

Figure 8.1. The 3 panels show the distribution of the number of tasks performed below the 3 criterions p16, p5, and p1, from top to bottom.

8.3.   Visual Deficits in ADS

Subitizing may be impaired in ADS-children also, because among the brain structures affected in ADS there may be also those, which are involved in subitizing. ADS-children with a complete positive diagnosis of dyslexia have been excluded from this analysis. We

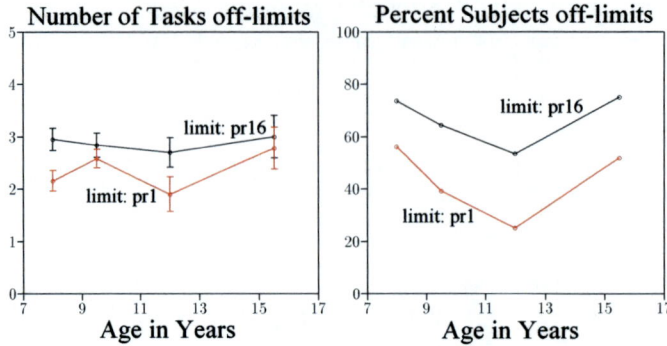

Figure 8.2. The left panel shows the number of tasks performed below the criterion of of p16 and p1 as a function of age. The corresponding percentage of subjects with ADS are shown by the right panel.

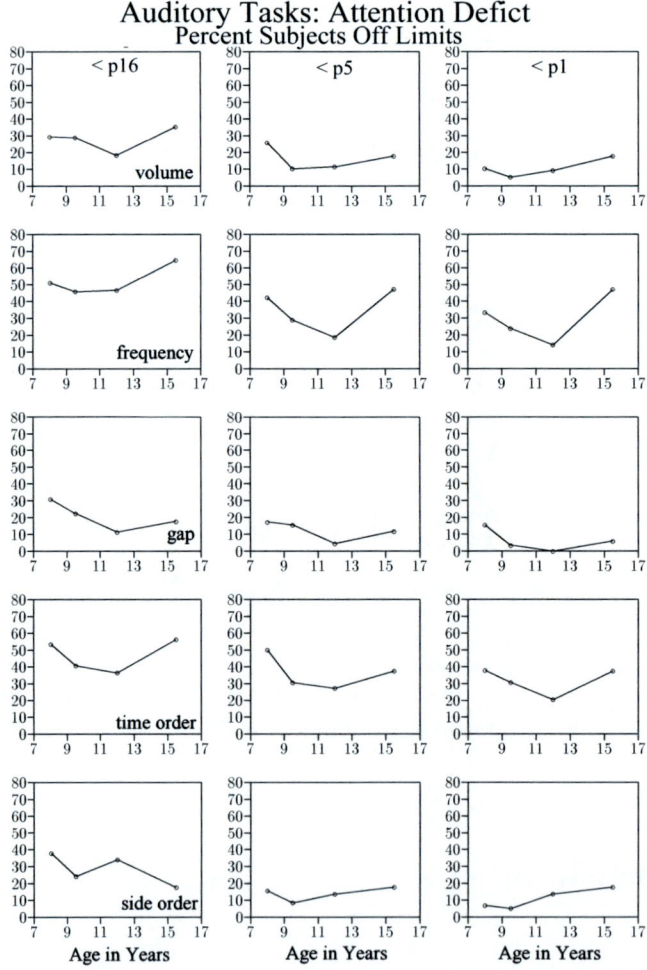

Figure 8.3. The graphs show the number of subjects failing to reach the criterion of p16, p5, or p1 (columns) for each of the 5 auditory tasks (rows). Note the differences between the curves when comparing the different tasks.

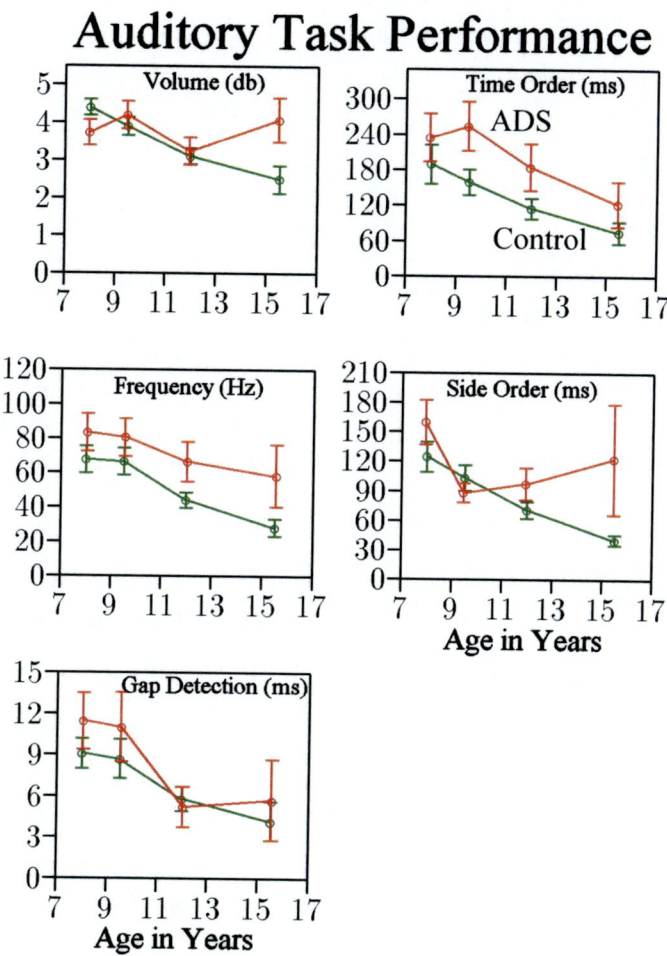

Figure 8.4. The panels show the pair of age curves of the mean of the threshold values obtained from ADS children in the 5 tasks.

look at the basic response time and at the combined variable of the effective recognition speed.

The pair of curves in Fig. 8.5 shows the differences in the basic response time at the left side and the percentage of children performing below the p16 limit at the right side. Clearly the children with attention deficits are slower at all ages. The percentage of affected children increases from about 30% to about 66%.

The Fig 8.6 shows the pair of age curves and the percentage of ADS-children performing below the p16 criterion. The figure shows the deficits at all ages. The percentage of children off-limits increases from 50% to about 75%.

While the differences in the basic response time are moderate and the percent number of affected children is not strikingly high, the discrepancy for item numbers of 4 and above is relatively strong and the percent number of affected children is considerably higher.

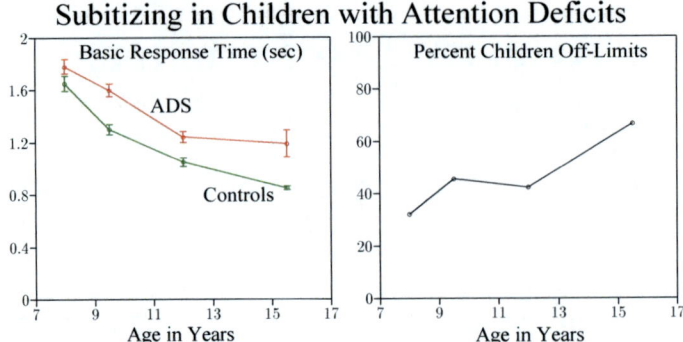

Figure 8.5. The curves at the left side show the differences between controls and children with attention deficits with respect to the basic response time in the task of subitizing. The right diagram shows the percentages of children performing below the p16 criterion.

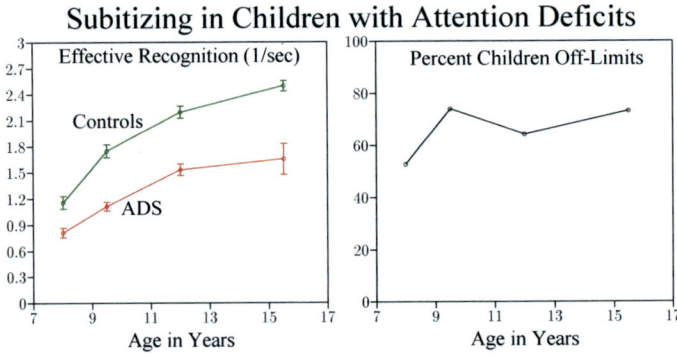

Figure 8.6. The figure shows the age development of the effective recognition in control subjects and in ADS-children. The right side shows the percentage of ADS-children performing below the p16 limit

8.4.   Saccade Control in ADS

As in the case of dyslexia we describe below the results of the pro- and antisaccade tasks using the variables defined in the first part.

The Fig. 8.7 shows the reaction times of prosaccades (overlap condition) and antisaccades (gap condition). Clearly, the reaction times of the prosaccades are slower for ADS children at all ages. The reaction times of the antisaccades on the other hand are about the same for the controls and the children with ADS, but only for the two younger age groups. For the older groups the reaction times of the correct antisaccades are significantly slower in the affected children. This result is very similar to that of dyslexia.

The errors made in the antisaccade task are also prosaccades. We therefore may expect that this variable differentiates the dyslexics from the controls. In fact, Fig. 8.8 shows at the

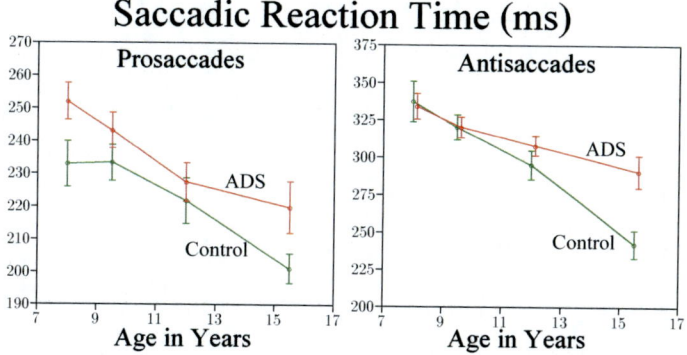

Figure 8.7. The panels show the age curves of the saccadic reaction time of prosaccades and antisaccades of the control subjects and the subjects with ADS.

left the pair of age curves. A real discrepancy is seen only for the oldest age group. The time for the corrective saccades differ at all ages but not for the youngest group.

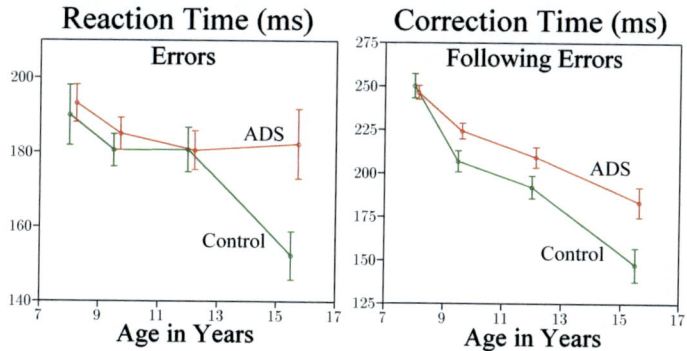

Figure 8.8. The panels show at the left side the age curves for the reaction times of the errors and at the right side the correction times for control subjects and subjects with ADS.

In the antisaccade task with gap conditions children make surprisingly many errors by looking to the stimulus first. This movement was not planned and not required. By contrast, this movement was forbidden. Children with ADS make more errors as can be seen in the Fig. 8.9. The left side shows the error rates in percent. Both groups start with high error rates at the age of 7 to 8 years. The error rate decreases for the controls and much more slowly for the children with ADS. The percentage of corrections are about the same for the two younger age groups but for the older groups developmental deficits can be seen: The correction rate increases with age in both groups, but the children with ADS are leaving more errors uncorrected.

The combined variable of misses:

$$pmis = perr \cdot (100 - pcor)/100$$

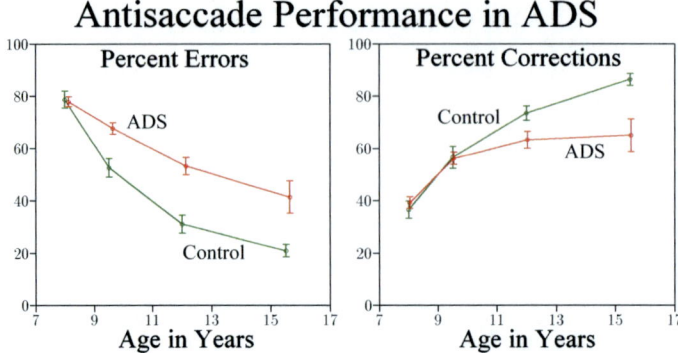

Figure 8.9. The age curves of the error rates (left) and the correction rates (right) are shown for the control group and the group of subjects with ADS.

(indicating in how many trials from all trials the initial error is not corrected) is shown in the Fig. 8.10 by the pair of curves at the left side. Both groups start with high percentages of misses. With increasing age the curves diverge: more and more children with ADS could not reach the range of the controls (p16 limit) as can be seen at the right side of the figure. In fact, the percentage of children with ADS reaches about 70%.

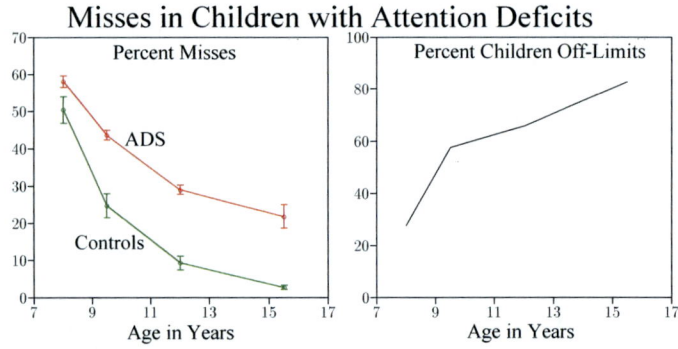

Figure 8.10. The figure shows the development of the rate of misses (left side) and the percentage of subjects with ADS, who failed the p16 limit of the controls.

When doing the same analysis for the reaction time of the correct antisaccades only small differences are obtained between the groups. In particular, the younger children with ADS have as fast reaction times as the age matched controls. Only the older groups are slower.

The symptoms of ADS-children are often treated by a medication with Methylphenidat (Ritalin). This drug effects also the generation of saccadic eye movements [Klein et al. 2002]. We will present the data in the part Training showing an improvement in saccade control by the medication.

Chapter 9

General Learning Deficits

Summary

This chapter is devoted to children with learning deficits, which could not be classified into one or the other "classic" groups dyslexia, dyscalculia, attention deficits disorder. These children are not mentally ill or retarded, but their general intelligence may be somewhat below the normal range. The results will show that in this group most children are severely impaired in most of the auditory, visual, and optomotor domains examined.

9.1. Characteristics of the Learning Deficits

One tries to characterize children with learning problems as dyslexic or children with ADHS or as belonging to other more or less clearly defined groups. This might be necessary for scientific purposes when trying to understand the neurobiological background of a certain kind of learning deficit. However, there are many children, which also suffer from learning problems at school, who cannot be classified into one of the well known groups. Mostly these children exhibit deficits in their general intelligence (IQ below 80) and therefore one believes that they do not learn how to write and to read, because they are not learning anything else to a certain normal level. Yet, these children cannot be regarded as mentally retarded or ill.

This chapter is devoted to this "poorly" defined group of children. We want to know if these children also suffer from basic auditory or visual or saccade control deficits and how many of them are affected. To make the children comparable (not only in age) the participants were selected from a single school all getting their lessons from the same teachers and all living in the same school environment. These aspects become important, when we try to see whether or not these children can improve their deficits by training and whether or not a successful training would transfer to their performance at school.

9.2. General Learning Deficits: Auditory Functions

The 5 diagnostic auditory tasks were completed by 49 children in the age range of 9 to 15 years. As in the other groups treated so far the children were unable do one or the other task better than by chance or they failed the p16 criterion. The Fig. 9.1 shows in the top diagram the distribution of the number of tasks that were performed below the p16 criterion. Note, that not a single child was able to do all 5 tasks within the range of the controls. The vast majority (90%) failed in 1 or more tasks and 9/49 (18%) even failed in all 5 tasks. Most subjects (65%) failed in 3 or more tasks. About 80% were at least impaired in all 5 tasks. When the p5 criterion was used still 98% were impaired in 1 or more tasks. With the p1 criterion still 90% failed in 1 or more tasks (see bottom diagram of the figure). In other words: the total group was severely impaired in the domain of auditory discrimination.

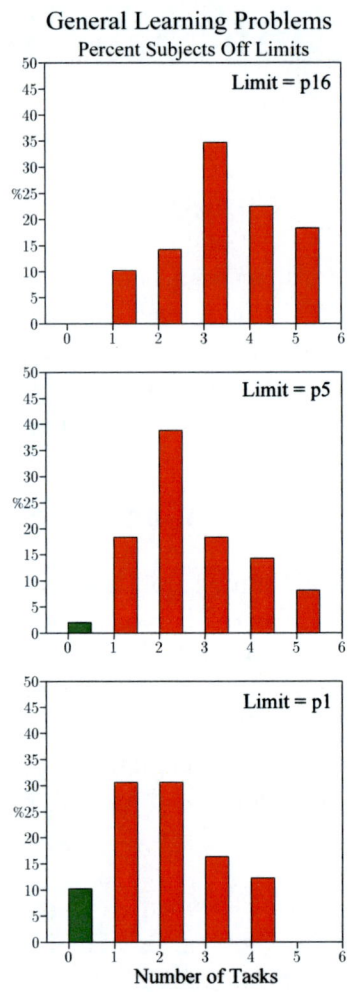

Figure 9.1. The figure shows the number of auditory tasks, in which the criterion of p16, p5, or p1 were failed. Note, that these numbers remain high when the criterion was changed from p16 to p1.

The effect of age on the number of failed tasks is shown by the Fig. 9.2 for the p16 and the p1 criterion. The left side shows the mean number of failed tasks and the right side depicts the percentage of subjects failing these tasks. Note the high number of tasks failing the p16 criterion (left side, upper curve). Of course, all subjects (100%) contribute, because we already know, that not a single subject was able to do all tasks within the limit of p16. Using the extremely strong criterion pf p1 reveals that on average still 2 tasks failed the criterion and that 80% or more achieved no better results.

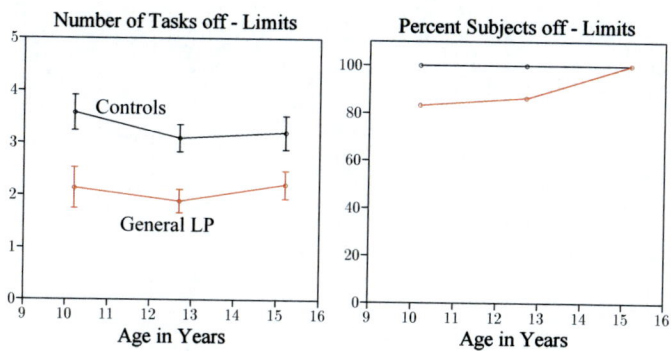

Figure 9.2. The figure shows the age curve of the number of tasks failing the p16 and the p1 criterion (left side). The panel at the right shows, that all subjects failed the p16 limit and 80% or more failed even the p1 criterion.

The next step leads us to consider the different task. We use the 3 criterions of p16, p5, and p1 and we count how many subjects failed the task by these criterions as a function of age. The Fig. 9.3 shows the graphs. The most dramatic deficits are obtained in the frequency domain and in the time order task. Stronger criterion hardly reduce the percentage of children performing the tasks below the criterion. Time order seems to be an especially big problem in these children.

Finally, we look at the age curves of the mean values in comparison to those of the age matched controls. The Fig. 9.4 presents these pairs of curves for all 5 tasks. What could be expected from the above analysis large deviations from the control curves can be seen. Again the frequency domain and the time order domain exhibit large deficits. Among the test subjects there exists a large scatter can be seen also.

9.3. General Learning Deficitis: Subitizing

The variables describing subitizing are evaluated for the group of children with general learning problems and illustrated below by the following figures. First we look at the basic response time needed to give the correct response for 1 item (out of the nine possible items). The Fig. 9.5 shows the pair of curves from the control subjects and the test subjects. The deficits are evident at all ages, the response times of the test children are slower by about 0.3 sec. At the right side one also sees the percentage of subjects among the test subjects who failed the p16 criterion. While the two younger groups contain only 40% subjects off-limits, while the oldest group contains more than 60%, even though the mean values are closer together. The reason is the smaller scatter in the test as well as in the control group.

The time per item and the percentage of correct responses from item number between 4 and 8 are illustrated by the pairs of curves in Fig. 9.6. The extra time for each item is clearly

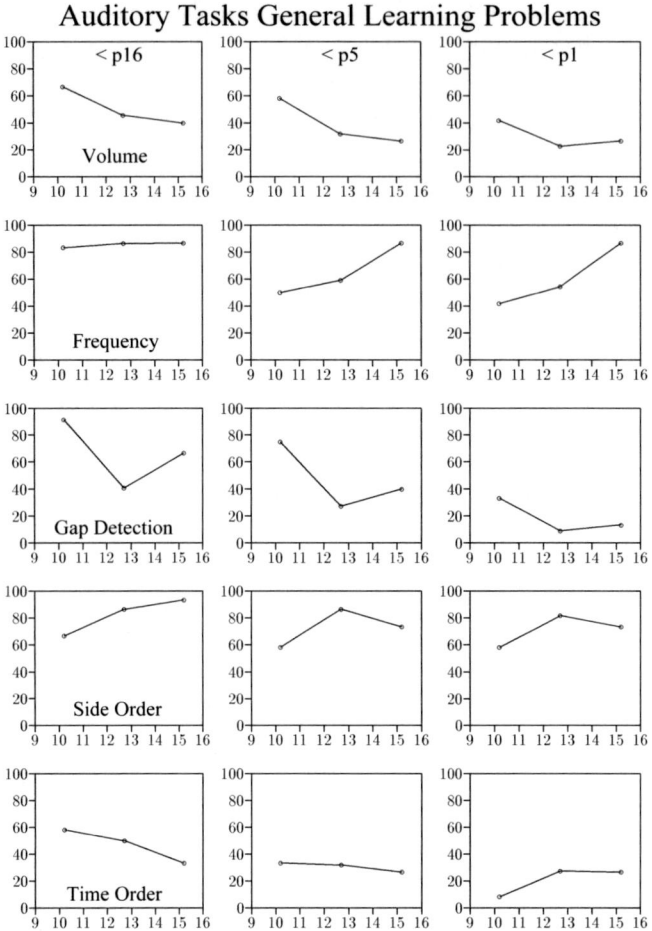

Figure 9.3. The number of tasks failing the criterion p16, p5, or p1 (from left to right) are shown for all 5 auditory domains (from top to bottom) as a function of age.

longer for the test subjects and the percentage of correct responses is way below the values reached by the controls. Correspondingly, the percentage of subject off-limits is extremely high (80 to 100%) as can be seen in the lower right panel. On the other hand, the criterion of p16 was missed by only half of the test subjects when considering the time per item as variable (lower left panel).

The combined variable defined as the effective recognition speed is shown in Fig. 9.7 together with the percentage of subjects failing the p16 criterion. Because both variables, the time per item and the percentage of correct responses, differentiate the test from the control group at all ages it does not come as a surprise, that the effective recognition speed exhibits also large deficits. All members of the youngest group and more than 70% of the other age groups failed the criterion of p16.

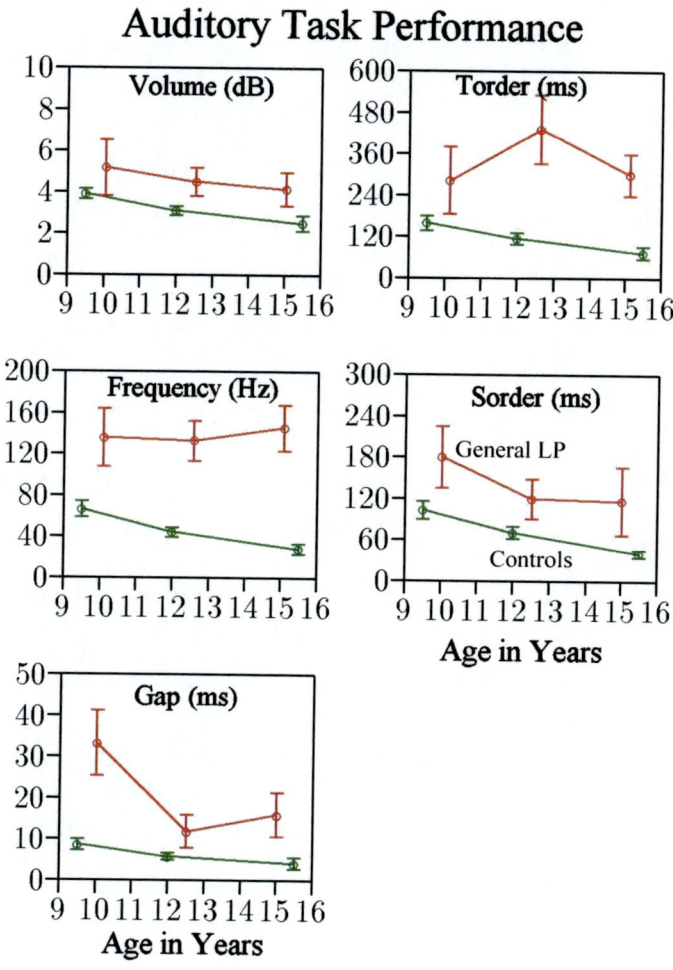

Figure 9.4. The figure shows the age curves of the 5 different tasks in comparison to those of the age matched controls. Note, that different bin widths for the age groups have been used for this figure.

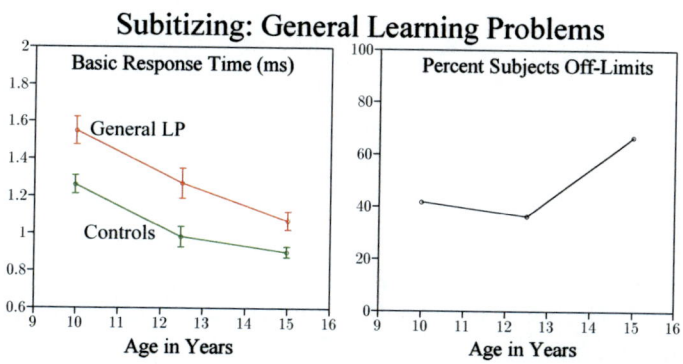

Figure 9.5. The graph shows the age curves of the basic response time obtained from the control and the test children. The right side shows the percentage of test children, who failed the p16 criterion.

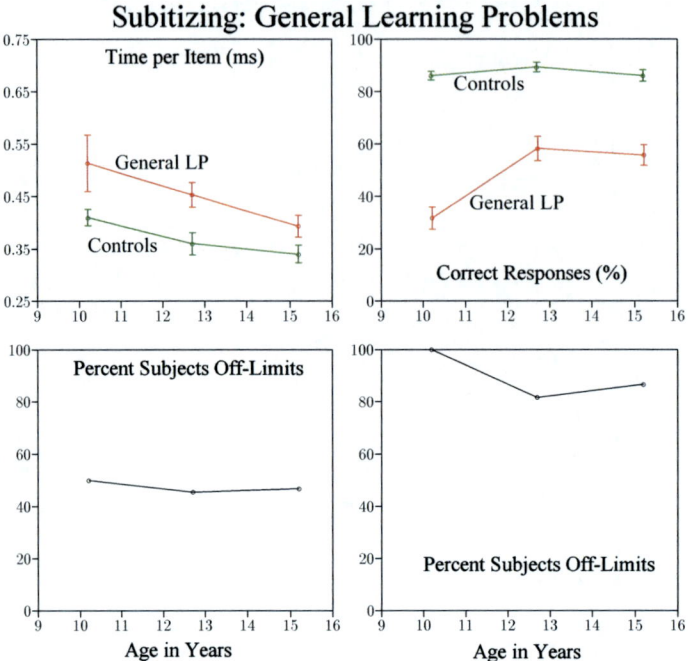

Figure 9.6. The age curves of the time per item (left side) and the percentage of correct responses (right side) shows the differences between the control and the test group of children with general learning problems. The lower panels show the percentage of subjects failing the p16 criterion.

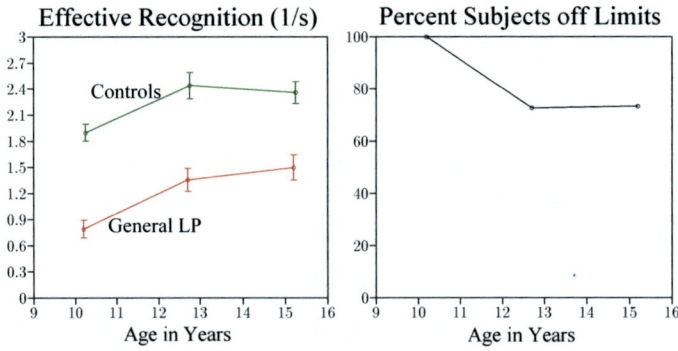

Figure 9.7. The left panel shows the effective recognition speed as a function of age for the group of controls and the group of test subjects. The right panel show the percentage of subjects failing the p16 criterion.

In conclusion, we have seen that the deficits in subitizing are relatively moderate if one considers the time domain (response time). The deficits are more severe when considering the correctness of the responses. In fact, none of the previously examined groups of subjects were as severely affected as the group with general learning problems.

9.4. General Learning Deficits: Saccade Control

The deficits in saccade control of this group of subjects is characterized by the pairs of age curves and the corresponding percentages of subjects failing the 16p criterion (off-limits). First we look at the reaction times of pro- and the antisaccades as shown in Fig. 9.8. Interestingly, in this group the reaction times of the prosaccades are slower in the test than in the control group. (Dyslexics and dycalculics did not exhibit systematic differences). Thirty to 40% of the subjects are affected. The reaction times of the antisaccades exhibit great scatter within the youngest test group and the mean value is faster in the youngest test as compared with the youngest control group. Correspondingly, only 20% and less fall outside the normal range. Only the oldest group was affected in about 50% of the cases.

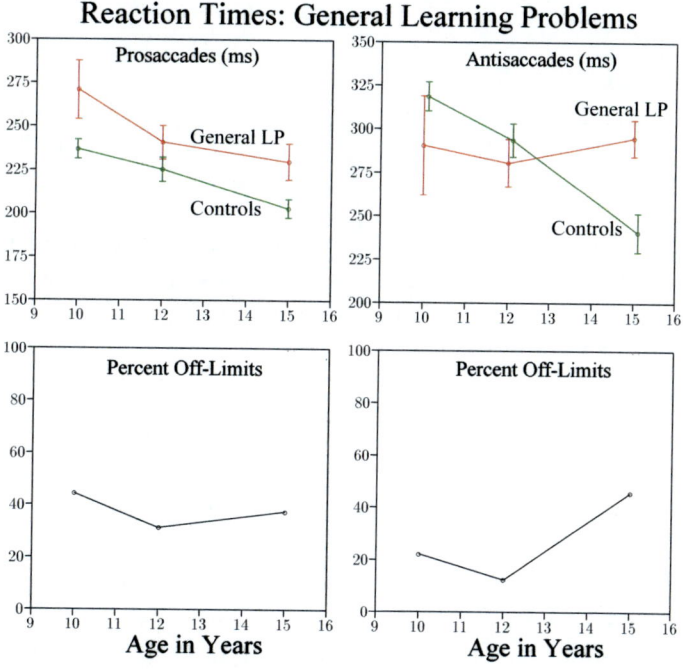

Figure 9.8. The pairs of curves show the development of the reaction times of the prosaccades (left) and the antisaccades of the test and the control children.

From the antisaccade task the percentage of errors and the percentage of misses are shown by the Fig. 9.9. In both variables many of the test children failed the p16 criterion. In fact, these percentages (40 to 90% and 80%, respectively) are the highest off-limit values we have encountered throughout all of the groups described in this book. This may be taken as an indication, that their frontal brain control of saccades is suffering considerably from a developmental deficit.

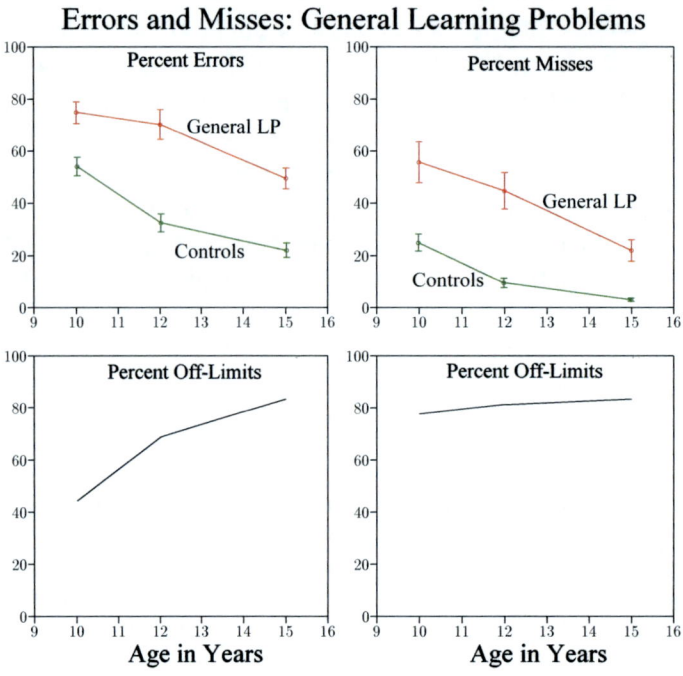

Figure 9.9. The figure shows the percentages of errors (left) and the percentages of misses as a function of age for both groups. The large deviations in both pairs is evident. Below the percentage of off-limit subjects is shown for each age group. Note the extremely high off-limits values in percentage of misses.

The time for corrections and the percentages of corrective saccades following the errors are also analysed. The Fig. 9.10 shows the corresponding pairs of curves. As with the reaction times of the antisaccades, the correction times are only slightly affected, but the percentages of corrective saccades are strongly affected at all ages. This difference becomes definitely clear when looking at the percentage of affected subjects shown by the lower panels of the figure: while the correction time hardly reaches 50%, the percentage of corrections reach values between 50 and 80%.

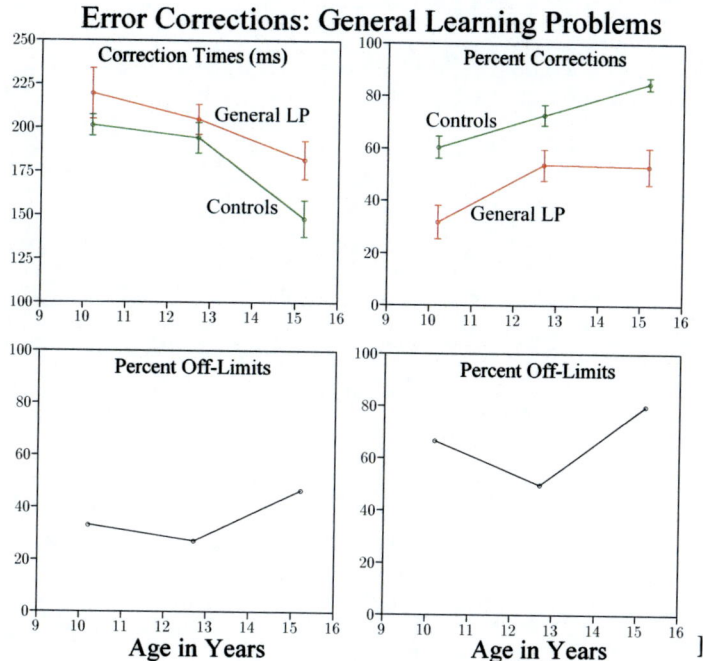

Figure 9.10. The pairs of curves show the development of the correction time (left) and the percentage of the corrective saccades. While the differences in correction time are not so dramatic the percentages of corrections differ considerably.

PART III

TRAINING

Summary

This part deals with the possibility to help those children, who failed one or all tests. The strategy is training, i. e. repetition of the performance of tasks, which challenge the brain specifically in the domain exhibiting the deficit. The methods are described, the results of the training presented and compared with the pre-training values. We will see, that the training methods improve the performance of the tasks systematically, but not all subjects were successful. The percentages of successful subjects are calculated and presented as a function of age.

Chapter 10

Principles of Learning and Training

Summary

This chapter deals with learning by training. The neural basis will be briefly explained and the features of the training will be described in detail. The principle of the training follows the rule: repeat the tasks that you failed day by day for several weeks. The various groups of subjects will not be treated separately. For the children with attention deficits we consider also the effects of medication and the training effects with and without medication. The training effects for children with general learning problems are also treated separately.

At birth the human brain controls only a few functions. Even basic functions like audition or vision have to be established during the following months and years. We have already seen that many of these functions keep improving until the adulthood.

This kind of plasticity makes it possible for our brains to learn specific functions at different ages. While many learning processes – like learning to speak a language – should begin relatively early, i.e. during the first 2 years, other learning processes may start later and take longer. The plasticity of our brains also makes it possible to improve functions, which are not developed well enough at a certain age (developmental deficits). Even in cases where the developmental deficit appears to be the consequence of genetic preconditions, the nervous system may find ways for compensation to perform certain tasks, for which it was not quite prepared genetically. The plasticity of the brain serves as a basis for the training procedures described in the following chapters. We apply to the sensory systems similar principles as for the training of the motor system.

10.1. A Neurophysiological Principle of Learning

It has been known for many hundreds of years (or even longer) that the repetition of certain specified acts results in an improvement of the performance of this act. For example, in sports or in musical education (playing an instrument) the principle of learning by repetition is used up to now as a regular and accepted method of learning. In school the principle of repetition has become somehow unpopular and boring.

From neurophysiology we know that the activity of subgroups of neurons depend strongly on the effectiveness of the synapses connecting the neurons within the group. If one assumes, that the performance of a task becomes possible by several groups of neurons (forming a network) acting both in serial and parallel pathways it follows that the quality of the performance is a function of the synapses in this network. Consequently, in order to learn the performance of a task, one wants the synapses to function as properly as possible. The question of learning therefore is reduced to the question of synaptic transfer of nervous impulses. How do we increase the reliability of synaptic transfer?

In 1949 Hebb formulated his idea of learning by the claim, that the probability of a successful transfer of an impulse from the presynaptic neuron to the postsynaptic neuron is increased by successful transfer of previous impulses [Hebb, 1949]. This neural principle of learning is illustrated by the Fig. 10.1. It shows schematically 2 neurons and the synapse, i.e. the connection of neuron 1 with neuron 2. Two impulses arrive at the synapse but only one is transferred to neuron 2. In order to challenge the synaptic transfer in those groups of neurons involved in the performance of a task, it is evident from Hebb's principle, that repetition of this task is the most favorable method to learn it. Learning by repetition is called training.

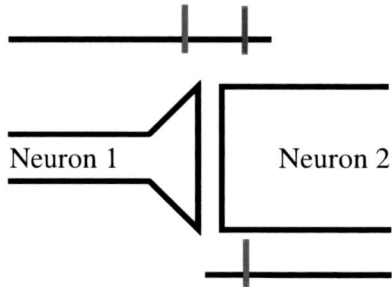

Figure 10.1. Schematic drawing of a synapse with the pre- and postsynaptic membranes. From 2 nerve impulses arriving from neuron 1 at the synapse. Only one is successfully transferred to neuron 2. The goal is to increase the chances of any impulse to cross the synaptic cleft between the two neurons.

From these considerations one can already see that the immediate goal of the training is NOT to improve reading or writing, but to improve those functions, that exhibit the developmental deficit. As an example, the Fig. 10.2 shows the age curve of a group of dyslexics and the corresponding age curve of the control group. The ultimate goal of the training is to move a member of the dyslexic group as quickly as possible into or close to the range of the controls. This effect is graphically indicated by the short black line in the figure. Training may result in partial improvement, because after the training the variable under consideration still does not fall within the range of the control group.

10.2.   Features of the Training

A serious training based on the neural principle of learning should have certain features. It should be:

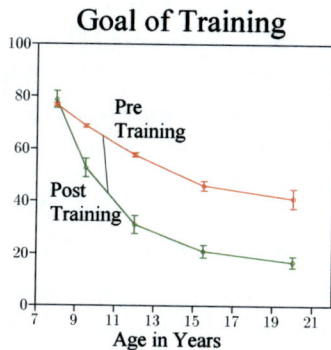

Figure 10.2. The short black line from the upper curve to the lower curve indicates the goal of the training: move the subject from the group with deficits into the range of the control group.

- **Evaluated:** The exact features of the various training procedures have to be found by experiences from earlier experimental work in the field. In the case of saccade control, it has been noticed many years ago, that daily practice can change the performance of saccade tasks [Fischer and Ramsperger, 1986]. Not all the single steps of evaluation of this training are described here, because not all contribute to the understanding of the training, but they are published and the reader can look up all the details in [Fischer and Hartnegg, 2000]. As a general rule, it was clear though, that a typical training period would last weeks, not just a few days or several months. It was also planned all from the beginning that the training would have to been done every day. It became clear after some experience, that age may play a role in selecting the time period for a certain training: older children need more training time as compared with younger children.

- **Selective:** The training must be as specific as possible to ensure that always the same synapses are challenged. From what we have seen already this means that the diagnostic tasks should be the ones used for the training. In the cases discussed here the tasks for training do not require any language processing. This makes an international application of the training possible.

- **Adaptive:** Since the subjects, who are supposed to do the training, have failed to pass the task during the examination, one needs very easy versions of the task to begin the training. As the child succeeds to do the easy version, the difficulty should be increased (automatically and in small steps) to challenge the brain function just at its present limit.

- **Controlled:** Many things can go wrong during the training period. The child may not like the training at all and tries to circumvent the daily sessions in one or the other way. Illness or other obligations may keep the child from doing the training regularly everyday. Depending on the general situation, the daily accomplishments are different from day to day. The child is unable to improve the performance and does not exceed a certain level of difficulty. All these possible reasons for an unsuccessful training must be taken into account. A training protocol after the training period helps to see the accomplishments of the child. Of course, the child and its family want to know also, how the training developed. Also, the fact that a training protocol is kept in the instrument and send to the parents may also contribute to the compliance.

In the following chapters we will not differentiate between groups of subjects with different problems like dyslexia or attention deficits or dyscalculia. Preliminary analysis of the data have shown, that the training is about as effective in one group as in the other. The really important factor for the effectiveness is compliance, i.e. the rules established for the performance of the training must be followed closely.

In the case of attention deficits the role of medication, that is often used for children, on the training will be considered separately.

Chapter 11

Training of Auditory Functions

Summary

To improve the auditory functions, in which the subjects did not reach the p16 (or even failed the p1) criterion, the same tasks were used for training as for the examination. However, different levels of difficulty were available. The training started by using a very easy version. The difficulty was increased in parallel with the child's accomplishments. The principle is to challenge the same auditory sub-functions thereby requiring those synapses which are involved in solving the task to become active in each single trial of the task. The training and the results have been described in detail earlier [Schäffler et al. 2004]

11.1. Procedure of Auditory Training

The training was done at home using a custom made hand held device with a built-in response keypad. A small LCD screen provided feedback after each trial during the training sessions. Each child performed the training of a given task for 10 days with a daily session lasting about 10-15 minutes. The tasks begin in an easy version on day one. The difficulty was increased as the child's performance increased. Only one task per day was required for training. The tasks were always performed in the order as described previuosly. The details of the training procedures were stored by the instruments and allowed a later analysis of the training. The corresponding training protocol was analysed and sent to the parents. It allowed to control for compliance. If the training was not successful because of poor compliance or because of only small or unstable training effects, the training of the corresponding task was continued for another 10 days. Three criteria were used: (i) percentile of 20 was not yet reached (low achievement criterion), (ii) task performance in one of the last two sessions was far below that of a previous session (stability criterion), (iii) performance level was increasing too slowly and had thus not yet reached a plateau (completeness criterion). If compliance was correct but the training was still not successful after the second training period, the training was not or no longer repeated.

The percentiles before (pre) and after (post) the training were determined for each subject and the respective mean values of all children were calculated. The percentages of subjects who performed above the 20-percentile after the training were determined as success rates and their respective mean percentiles were also calculated.

11.2.   Reduction of Unsolved Tasks

A first global impression of the effects of the auditory training is achieved by counting the number of the auditory tasks that were unsolved before and after the training. The Fig. 11.1 shows the distributions before (left) and after the training (right). The improvement is seen by an increase of the heights of columns towards the left and decrease towards the right in the right diagram as compared with the left.

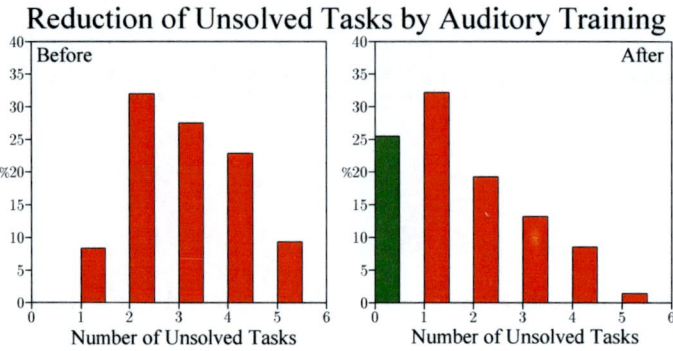

Figure 11.1. The panels show the distributions of unsolved tasks before (left) and after (right) the training. By definition the number of unsolved tasks before the training was never zero. The case of all tasks being missed occurred in less than 10%. After the training 25% of the subjects were able to solve all tasks and more than 30% missed only 1 of the tasks.

The distribution of the differences between the pre-and post training values are shown in the left side diagram of Fig. 11.2. Only a vanishing small percentage of values are negative (no improvement). About 20% of the subjects were unable to reduce the number of unsolved tasks. About 64% reduced the number of unsolved tasks by 1 or 2. The mean values before and after the training were significantly different by 1.42 tasks. The right panel shows the age dependence of reduction of number of unsolved tasks: older subjects profit more than younger subjects.

11.3.   Task Specific Improvement

The effect of the training was different when the individual auditory tasks were compared. The Fig. 11.3 shows the mean values before and after the training for several variables.
     The left panel shows the percentage of subjects who reached p20 (not only p16) after the training, indicating that they were successful in reaching the normal range of the controls. For the first 3 tasks this percentage is between 60 and 80%. In the 4th task (time order) only 40% were successful, and in the last task (side order) only 10% of the subjects reached the criterion of p20. In other words: the last task could not be learned by 90% of the subjects. We will see later, that despite this failure in task 5 (side order), the subjects profit in spelling.
     The right part of the Fig. 11.3 depicts 3 more variables describing the effects of the training for each task domain. The mean value of the pre-training rank order (pr-values) is

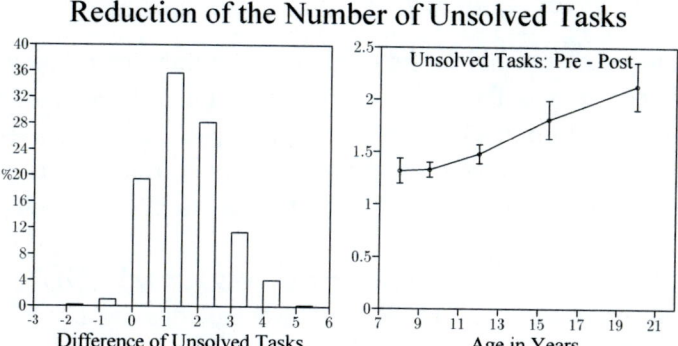

Figure 11.2. The figure shows at the left side the distribution of the change in the number of unsolved tasks from before to after training. The right side shows the mean values of the differences as a function of age: the older subjects profit more than the younger subjects.

depicted by the height of the first of each of the triplet of columns. These values were all below 5, because only those subjects participated in the training who failed the p16 criterion. The second column depicts the post-training mean value of the rank order. The heights of the columns show that the mean rank order values reach high values for the domains of volume, frequency, and gap detection. For time order the mean value is considerably smaller, and for the side order domain it did not even reach the p16 limit.

The last of the 3 columns in each triplet depict the mean values when only the successful subjects were included. All of these columns reach values above p30. This shows that the subjects who compled the training successfully, reach very high pr values, many of them even above the mean of the age matched controls (p50). Note, that those few subjects, who reached the criterion of p20 in the side order task, also reached high pr values. This means: once the training is successful, its result is excellent.

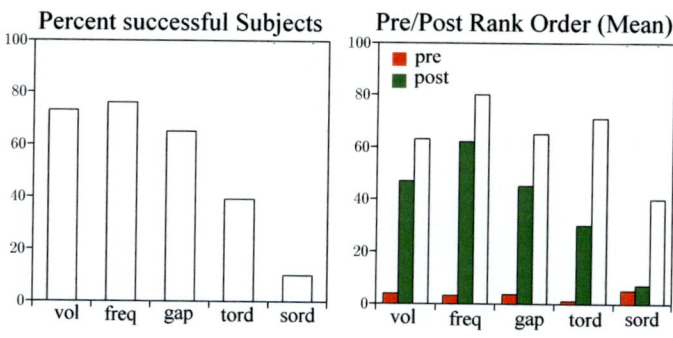

Figure 11.3. The figure shows on the left the success rates in percent for the 5 auditory tasks. The diagram at the right shows the mean rank orders before and after the training. The white columns indicate the mean values, when only the successful subjects having reached p20 or better were included.

11.3.1.   Specificity of the Training

The rank order reached by the end of the training may exhibit correlations indicating that if one task was learned, so was the other. The correlation matrix consists of 10 correlation coefficients. They range from 0.17 to 0.61 and some reached significance because of the large number of subjects. However, the prediction of a learning effect in one task from knowing whether another task was learned, was too low to allow skipping the training of task 1.

Interestingly, the frequency discrimination task was learned easily and by many subjects, while the time order task was difficult to learn and considerably fewer subjects succeeded. This is another indication for the notion, that the auditory domains of frequency discrimination and time order are functionally independent.

More evidence in favour of the independence comes from the consideration of subjects who initially failed the p16 criterion in both the frequency and the time order domain. About 78% of this group reached the p20 criterion in the frequency domain, but only 36% reached the p20 criterion of the time order domain. Also, from those, who failed to learn the time order task, 63% succeeded in learning the frequency task. But from those, who failed to learn the frequency task, only 5% learned the time order task.

11.3.2.   Training Effects of Severely Impaired Subjects

In the part "Deficits" we considered the effect of the strength of the criterion (p16, p5, and p1) on the number of subjects failing it, and we noticed that especially for the frequency and time order domain, this number remained surprisingly high when the criterion was stronger. This means that in these domains many children had not just weak, but strong impairments. Now we want to know, how many of the severely impaired subjects (failing p1) were able to improve their auditory capacities, reaching p20 after the training. The Fig. 11.4 shows the answer. The percentage of successful subjects is about as high as with the criterion of p16 before and p20 after the training. This argues against the possibility, that severe cases do not profit from the training. It is the task which determines the success rate, not the amount of the impairment.

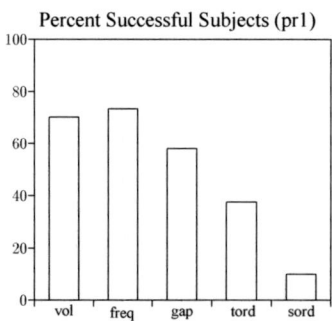

Figure 11.4.   The percentages of successful subjects starting with p1 (severely affected subjects) before and reaching p20 after the training. Severely impaired subjects were about as successful as others.

11.4. Auditory Training: General Learning Deficits

The group of children with general learning problems needs an extra analysis of its training data, because it might be that all or many of these children are unable to improve their perceptual performance at all. We first look at the number of unsolved auditory tasks before and after the training. The Fig. 11.5depicts the data before and after the training. Before the training none of the children were able to complete all 5 tasks above p16, only 5% and only 10% missed 1 or 2 of the tasks, respectively. Twenty percent could not do any of the tasks within the normal range. After the training 10% could do all tasks above p20, another 10% failed p20 in one of the 5 tasks. Only 5% were left who could not do any the tasks. The auditory training was successful in this group, but the amount of success was certainly smaller when compared with the group of dyslexic children described above.

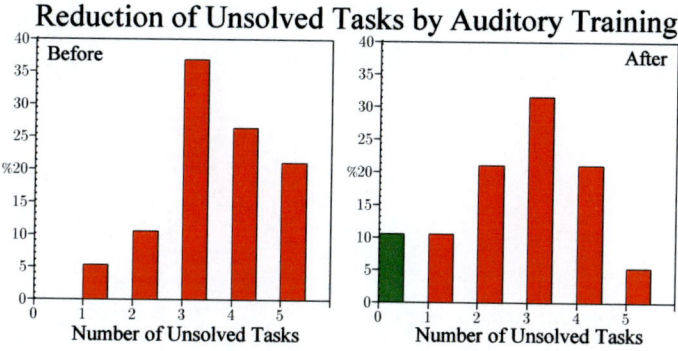

Figure 11.5. The figure shows the distribution of unsolved auditory tasks before and after the training of a group of children with general learning problems.

Parallel with this experimental group, there was also a waiting group in the study. The members of this group were also examined after the training period to see, whether the time elapsed during the training (normal age development of auditory system) or any other cir-cumstance (e.g. attend school) would also lead to an improvement in the auditory domain. The mean value of their number of unsolved tasks, however, remained the same as before. Only a few subjects reduced their number of tasks below p16.

As in the case of dyslexia, we also looked at the differences in success of the 5 tasks separately. The Fig. 11.6 shows at the left the percentage of children who reached the p20 criterion after training although failing to reach the p16 criterion in the initial examination. These percentages vary between 25 and 50 depending on the task. The frequency domain was the most difficult one: only 25% of the subjects could improve their performance. The right side shows the mean percentiles before the training (small columns) in comparison to the mean values after the training. All values were below 30 indicating that the group as a whole did not reach "normality". But if one looks at the successful subjects only (right) one sees that the mean percentiles are high (between 35 and 70), indicating that the auditory training works very well once it begins to work.

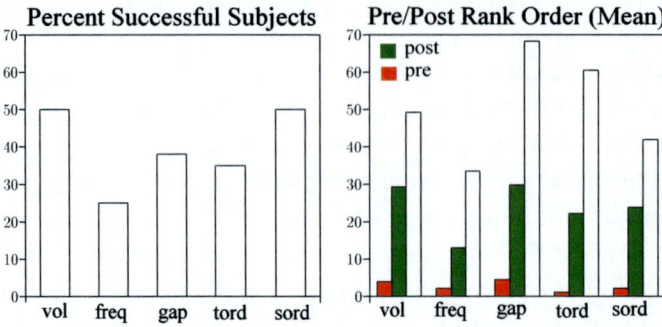

Figure 11.6. The figure shows the success of the auditory training in a group of children with general learning problems. The left side shows the success rate for each task. The right side shows the mean of the rank orders achieved before and after the training. The right column of each triplet shows the rank orders of the successful subjects.

Chapter 12

Training of Visual Functions

Summary

The philosophy of this chapter remains the same as before: repeat the execution of that task, in which the subject failed to reach the range of the age matched controls. From the 2 visual domains, dynamic vision and subitizing, we present only the training data of subitizing, because the tasks examining dynamic vision are used for the saccade training. As a rule the children reached very high scores even with increasingly high speeds of stimulus presentation, when they do the saccade training. Again, only one large group of subjects is considered, because there were no significant differences between the effects of the training for dyslexics, children with dyscalculia, or children with ADS

12.1. Procedure of the Training of Subitizing

The training task for subitizing is essentially the same as the one used for the examination with the following exceptions: because the children could not perform the test task according to their age control group the task was made very easy at the beginning of the training by presenting only item numbers up to 3 and by increasing the presentation time to 300 ms instead of 100 ms. As the child learned to perform this easy task, the difficulty was automatically gradually increased by allowing more item numbers and by decreasing the presentation time. There were several combinations of these two variables that determined the levels of difficulty. Each training session consisted of 140 trials and lasted about 10 to 20 minutes.

At the beginning of the training period of 20 days each subject had to perform the test task in its original form to record the pre-training performance. Similarly, a test session had to be repeated at the end of the training at day 21 to have the post-training values. All data were stored by the training instrument and were used to write a training protocol after the training device was returned to the lab.

The visual display, the keyboard, and data collection were all implemented in a small hand held instrument. The children were given one instrument each to practice the tasks one session every day.

12.2.   Changes in the Counting Process

To illustrate the effect of the training on the counting responses, the Fig. 12.1 shows a pair of curves of the percent of correct responses as a function of the number of items. While for item numbers below 4 there are no training effects visible (because the percentage is already close to 100 creating a sealing effect), the subjects made after the training smaller number of mistakes as the item number increases above 4.

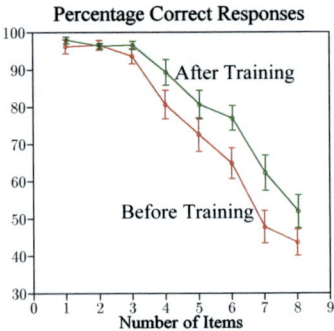

Figure 12.1.   The curves show the percentage of correct responses as a function of the number of items before and after the training. Note, that for item numbers 1 to 3 almost all responses were already correct before the training and no training effect can be seen. The children were 9-10 years old.

The responses times as a function of the number of items are shown in Fig. 12.2. The response times are decreased by the training even for item numbers below 3, indicating that subitizing was improved in the time domain, while the percentage of correct responses was close to 100% all from the beginning. This fact explains why we may believe, that subitizing of 1 to 3 items is well established at the age of 10 years. Subjectively, one does not notice the time gained by the training.

12.3.   Comparison of Pre- and Post-training Variables

As in the auditory domain we have to consider the different variables describing the task performance. The same analysis is carried out for the training effects in subitizing.

The pre- versus post-training value scatter plots are shown in Fig. 12.3. Data points on or close to the diagonal lines indicate no change of the values. The high correlations are mostly a consequence of the age dependence of each of the variables. Most of the response times of subitizing (1-3 items) fall below the line (speed increase) and most of the values of the effective recognition speed (ERS, 4-8 items) fall above the line (better recognition). Similarly, most of the values of the time per item (4-8 items) and of the percentage of correct responses (4-8 items) are improved. The scatter plot shows that the vast majority of children could improve their performance of the task of subitizing and counting by memory.

The data presented so far do not show, whether the children could improve more than one variable and the question of age dependence is left open, also.

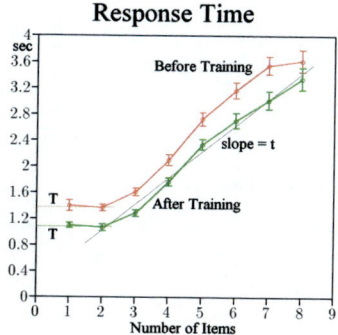

Figure 12.2. The figure shows the response time as a function of age before and after the training. Note that the response times for 1 to 3 items became shorter with training. A training effect for these item numbers is obtained in the response time, not in the correctness.

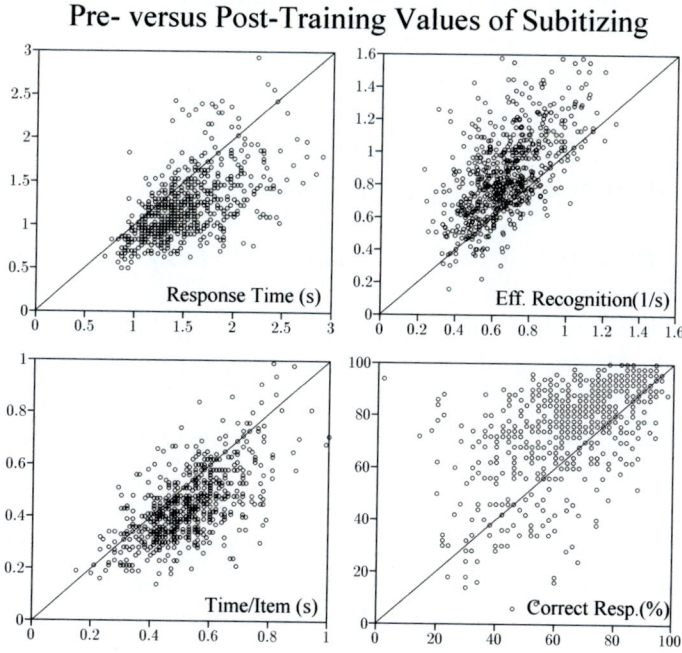

Figure 12.3. The figure shows the scatter plots of 4 variables describing the performance of the task of subitizing and counting by memory before and after training. Values on the diagonal line or close to it signal "no change" due to the training.

The age curves are shown for the response time and for the effective recognition speed, because only two of the four variables were found to be statistically independent. The two pairs of pre- and post-training age curves are shown in Fig. 12.4. Clearly the post-training curves are different for all age groups and for both variables. The amount of improvement given by the differences between the curves seems to be relatively constant. Only the ERS improvements seem to be larger as age increases.

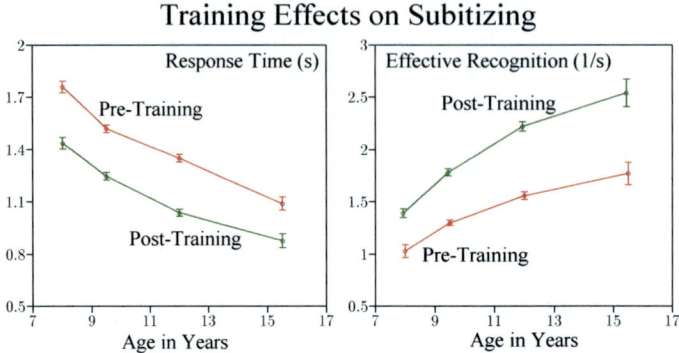

Figure 12.4. The panels show the pre- and post-training age curves of the response time and the effective recognition speed. Improvements can be seen in both variables and at all ages.

12.4.   Success of the Training of Subitizing

To really estimate the success of training it is not enough to show the displacements of the age curves. Some children may have accomplished high performance levels, while others remained at low levels or performed the task even worse than before the training.

The first and simplest way of estimating the success rate is to count the number of variables that have been improved by the training. Neither of the 2 variables were improved in only 5% of the cases, 1 of 2 variables were improved in 20% of the cases, and both variable were improved in 75% of the cases.

While the post-training curves are significantly different from the pre-training curves, the scatter plots have shown that not all subject could improve their performance in all variables to an extend that they could be regarded as members of the control group. In other words: We want to know, how many subjects could reach the normal range. By definition these "successful" subjects exhibit post-training values above the p16 limit of the control subjects of the same age.

Again we use the variables of the response time and the effective recognition speed. As a result it was found that these success rates showed only little variations with age and therefore we calculated the weighted mean value across age.

For the response time alone the success rate was 85%. For the effective recognition speed alone the success rate was 80%. If one requires that both variables reached the normal range after the training the combined success rate was 76%.

In addition, the success rate for the percentage of correct subitizing 1-3 items was found to be 84%. Remember that the response time for subitizing is very close to the response time for 1 item, T1. With T1 88% of the children reached the normal range. With T1

AND with the effective recognition speed 75% reached the normal range. In other words: subitizing can be improved by a very high percentage of subjects in the age range between 7 and 17 years. To illustrate this result graphically the Fig. 12.5 depicts the age curve of the control subjects and the age curves after the training of the response time and the effective recognition speed.

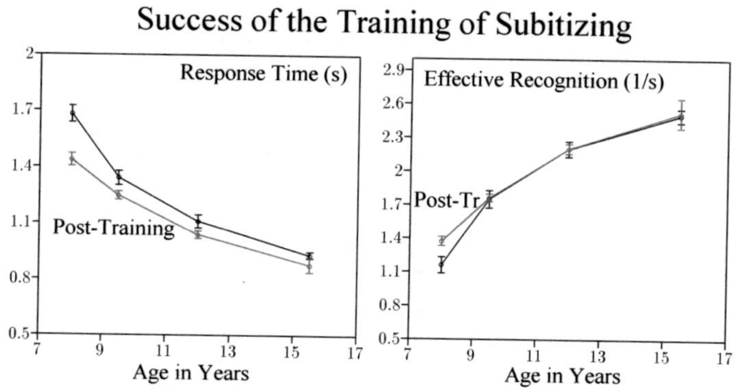

Figure 12.5. The figure shows age curves after the training in comparison with the age curves of the control subjects. The curves are almost identical indicating that high percentages of the trained subjects reached the range of the control subjects.

The curve of the post-training response times lies even slightly below that of the controls. One may expect, that in this case the success rate is 100%. Yet, it is in the order of 80% only. The reason is the scatter in the data of the post-training values: there are still subjects in the post-training group, who failed the criterion of the mean value + 1 standard deviation of the control subjects, even though the mean value of the post-training group is smaller ("better") than that of the control group.

Similar results are obtained, when the large group of children was subdivided into dyslexics and children with dyscalculia, etc. Therefore we are not showing these details here. There have been also subjects, who could not be classified into one of these groups. Yet, they profit from the training in about the same way. Children with severe learning problems of mostly unknown nature, however, will be considered separately below.

12.5.   Visual Training in Case of General Learning Deficits

The children with general learning problems were also given the visual training of subitizing and number counting by memory. The pre/post comparison is considered in this section.

All variables are analysed by calculating the relative improvement in percent. The waiting group also showed some improvement and therefore we show the percent difference values for both groups in Fig. 12.6. Each pair of columns shows one of the variables derived from the task of subitizing. The first and the second column in each pair indicates the improvement of the training and the waiting group, respectively.

The training group profits from the training in all 4 variables. The waiting group achieved better results also, but the amount of improvement is clearly smaller when compared with the group of children with dyscalculia. The basic reaction time (BRT) is short-ened in both groups, but the difference is larger in the training group. The time per item (TPI) is shorter in both groups and in this respect both groups are not significantly different. Note the large scatter in the effective recognition (ERS) and in the correctness of the responses (PCR).

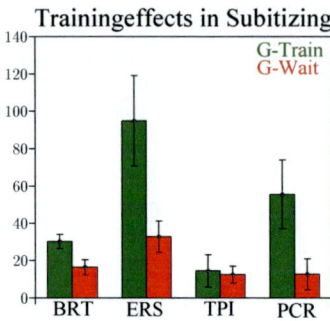

Figure 12.6. The columns indicate the percentage of the pre-post differences of the variables calculated for the training group (G-Train) and the waiting group (G-Wait). BRT: basic response time, ERS: effective recognition speed; TPI: time per item; PCR: percent correct responses.

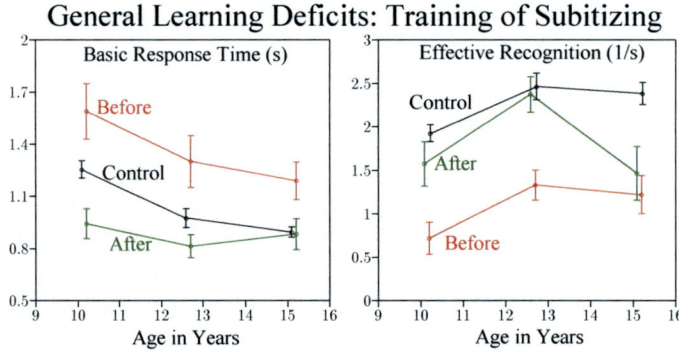

Figure 12.7. The three curves show the basic response time (left) and the effective recognition speed (right) before and after the training. The curves for the control groups are also shown.

The pre/post age curves of the basic response time and the effective recognition speed are shown in Fig. 12.7. The basic response time was improved after the training in all age groups. The age curve of the age matched control subjects is also shown in the figure. The two younger groups are after the training even faster than the control groups. The oldest group reaches the range of the control group.

The effective recognition speed shows a different picture: only the two younger age groups profit from the training and they reached the range of the controls. The oldest group did not profit from the training and thus failed to reach the control range by far.

The Fig. 12.6 has shown that the effective recognition and the percentage of correct responses profit most from the training. Yet, the detailed analysis shows, that only a relatively small number of the trained subjects reached the normal range. The conclusion therefore is that the improvement were considerably large, but for most subjects this improvements was not large enough to reach the normal range, especially in the oldest group. Perhaps one cannot expect higher profits from these children, because their general deficits prevent more progress.

Chapter 13

Training of Saccade Control and Fixation

Summary

This chapter deals with the training of saccade control. The elaboration of the individual training schedule rests on the comparison of the diagnostic values from the examination of saccade control with the data of a group of age matched control subjects. The effects of the training can only be seen by a repetition of the eye movement measurement. One has to observe the changes of all variables. These changes will be described and the post-training values will be compared with the data of the control subjects. The specificity of the training and the special effects on express saccade generation after the training will be presented on the basis of the data. Altogether we could re-examine 196 subjects. Details of the training have been described elsewhere [Fischer and Hartnegg, 2000]. We will see that the saccade training alone will not transfer to an improvement of fixation stability. A training with one eye covered, improves the binocular stability, but not the mono-stability.

13.1.   The Training Schedule and Procedure

We have to explain how the individual schedule for the training was found from the diagnostic values of the subjects. The corresponding training procedures will then be described.

At this point one has to remember the principles of the saccade control – the optomotor cycle – and its relation with the variables that are obtained from the analysis of the eye movements. To facilitate explanation of the elaboration of the training schedule on the basis of the diagnosis, we use the scheme of Fig. 13.1. It shows along the horizontal axis the strength of fixation and along the vertical axis the quality of the voluntary component of saccade control. The greater the distance from the centre – the normal range – the greater the deviation from normality.

The arrows pointing towards the centre represent the components of the training. The following variables are assigned to the axis:

- Weak fixation: the reaction times of the prosaccades are too short and/or the number of express saccades is too high. In this case a training of the fixation task is needed.

- Strong fixation: the reaction times of the prosaccades are too long and/or, their scatter is too large. A training of the saccade task is needed.

- Poor voluntary control: the percentages of errors are too high, the percentages of corrections are too small, the reaction times of the correct antisaccades are too long, the correction times are too long. A training of the antisaccade task is needed.

The box in the upper left corner of the Fig. 13.1 marks the position of the hypothetical data of a single subject in the scheme: the fixation is too weak and the voluntary component is too poor. The training of fixation and the training of the antisaccade task form the training schedule for this subject. Training of fixation moves the position to the right, training of the antitask moves it downwards. The combined effect of the training brings the subject's position closer to the centre region of normality as indicated by the vector sum of the two components. The rules for the elaboration of the training program rest on the concept of the optomotor cycle. Details of reading or spelling or arithmetic deficits do not play any role at this point.

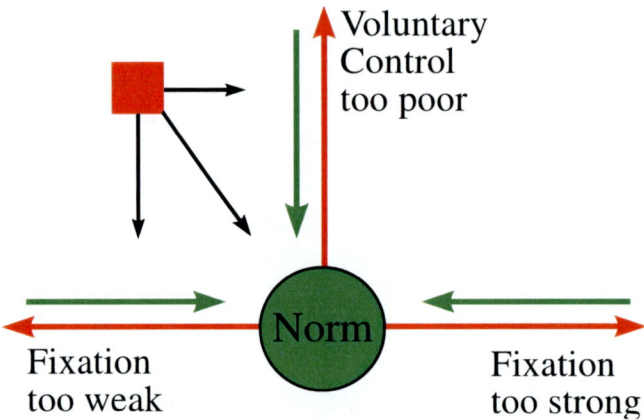

Figure 13.1. This scheme explains how the schedule for the training is elaborated from the diagnostic results of the analysis of the eye movements of a single subject. The diagnostic data determine the position of the subject (the square). To approach the region of the norm, the square should move to the right (train fixation) and down (train antitask).

The requirement of allowing an everyday training at home was solved by using the tasks of dynamic vision. First: these tasks could be easily implemented in a small hand held box given to the children for daily use at home. Second: the tasks challenge the three components of the optomotor system. The details have been described in the first part of the book. Briefly: the fixation task (F) requires the subject to maintain the direction of gaze at the centre of the display because this is the optimal strategy for solving the visual task of orientation detection. Similarly, the saccade task (S) requires timely saccades to the right or left, and the antitask (A) requires suppression of reflexes and antisaccades with respect to the distractor presented on one side, while the test stimulus occurs at the opposite side. In summary: the training of saccade and fixation control uses a visual trick: it gives a visual task which requires specific eye movement control to be solved correctly.

Using these 3 training components and the scheme of scheme of Fig. 13.1 the schedules for 196 subjects were found. The frequencies of the possible combinations are illustrated by Fig. 13.2. It shows only very few cases in which only one of the 3 tasks was used all below 8% (left triplet). The most common combination was fixation (F) & anti (A). This is clear from the diagnostic values, in which problems with the antisaccade task and weak fixation were encountered most commonly. Note, that the two tasks requiring saccades (the S and the A) were rarely used alone. The reason is, that in general the A-task was combined with the F-task in order to stabilize fixation to prepare the subject for the A-task. Remember: a successful antisaccade requires two sub-functions: suppression of prosaccades by stable fixation, and the ability to move the eyes to the opposite side.

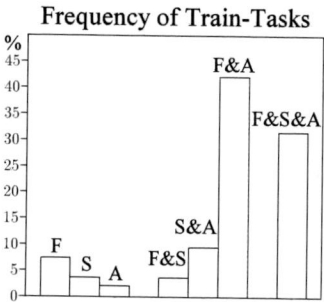

Figure 13.2. The graph shows the percentages of training tasks and combinations of them used for improvements of saccade control and fixation. F = Fixation task; S = Saccade Task; A = Antitask. Note: the A-task was included in almost all cases.

From earlier experience of studying training effects on saccade control it was known [Fischer and Ramsperger, 1986], that adult subjects could change their saccadic behavior within about 7 to 14 days of daily practice. Therefore, we decided to have the children practice each component by at least 7 -10 days. The fixation task, if necessary, was used first, followed by the saccade task (if necessary). The antitask was always trained at the end, such that the children could profit from the previous training when trying to learn the antitask. In this respect the training is a build-up training. Each daily training session consisted of 200 trials. A single training session lasted 7 to 15 minutes.

To facilitate the correct performance of the training tasks. they were given in an easy form at the beginning, becoming more difficult as the child's performance improved. The difficulty of the tasks were controlled by selecting the presentation time of the stimuli between 210 ms (easy) and 120 ms (most difficult). At the most difficult level the stimulus rotated so fast, that many parents were unable to do the task, because of their age. By training they could also learn it, but they would have to begin with an easier version. If one tries to do the task one gets an impression of what "fast vision" means.

13.2. Comparison of Single Values From Single Subjects

First we look at the data from single subject (11 y old dyslexic boy) before and after the training. His data are shown in Fig. 13.3. The left side depicts the pre-training data, the right side depicts the post-training data. He suffered from too slow reaction times and too large scatter in the prosaccade task with overlap conditions. The distribution of the reaction times indicate that he was unable to start his saccades in a defined period of time after stimulus onset. It looks as though the saccades were started at an almost random time. In the antisaccade task with gap conditions he made too many errors and he did not correct them often enough. The reaction times of the correct antisaccades were also scattered and so were the reaction times of the errors. The boy had to train the fixation task for 2 weeks, the saccade task for 1 week, and the antitask for 3 weeks.

The right side of the figure shows the data after the training. The reaction times of the prosaccades were shortened (from 292 ms to 217 ms) and very long reaction times (above 300 ms) were reduced in number. The reaction times of the antisaccades were considerably decreased (from 315 ms to 221 ms) and the percentage of correct responses was increased (from 38% to 53%). Finally, he also increased the percentage of corrections from 32% to 76%. The correction times remained about the same.

In principle, each of the different variables that could be extracted from the eye movement measurements could change due to the training. Therefore we look at the scatter plots of each variable before (x-) and after (y-) the training. Fig. 13.4 shows the 6 panels corresponding to the 6 pairs of pre-and post- variables.

The reaction times of the prosaccades do not show any systematic changes. The values are scattered above and below the diagonal. Only a few more points are found below the diagonal line than above it, indicating faster reaction times after as compared to before the training. Similarly, the reaction time of the errors (which are also prosaccades) show only a moderate systematic effect, but one can see, that there is a tendency for faster error reaction times after the training. This does not necessarily imply that the training does not affect the prosacade reaction times at all. We will discuss below how different components of the training influence the variables selectively.

The reaction time of the antisaccades and the correction times are clearly shortened by the training: most data points fall below the diagonal. Very clear displacements of the data points from the diagonal are also seen for the percentage of errors being reduced and for the percentage of corrections being increased.

13.3. Pre- and Post-training Age Curves

As in the chapters on deficits in saccade control we consider here the age curves of the different variables and their displacement due to the training. The Fig. 13.5 shows the mean values of the reaction times of the prosaccades with overlap conditions (left panel) and of the reaction times of antisaccades with gap conditions (right panel). While the reaction time curves are overlapping and close to each other with long error bars, the reaction time curves of the antisaccades are clearly separated.

The temporal aspect of antisaccade control is further characterized by the reaction time of the errors and by the correction time. The Fig. 13.6 shows the corresponding pairs of age curves. The errors exhibit large scatters, but they become shorter values by the training (remember, that the errors are also prosaccades). The correction times became faster at all ages. Large error bars occur for the oldest group because it contains fewer subjects. Since both variables together determine how long it takes the subject to reach the opposite side by

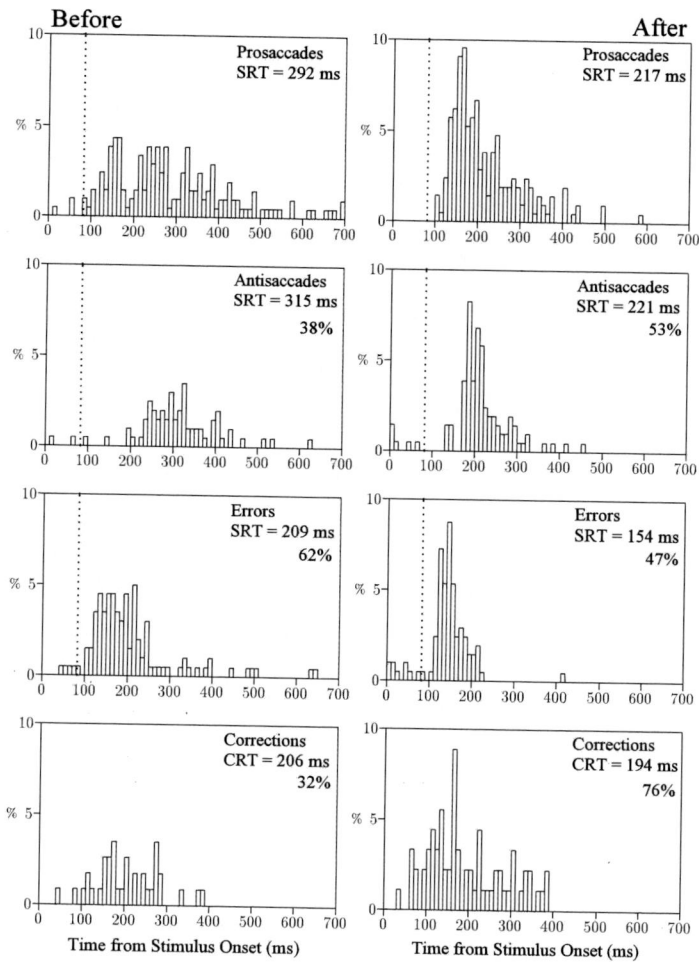

Figure 13.3. The figures shows the pre- and post-training data from a single dyslexic subject. The effects of the training can be seen by comparing the left with right panels. The changes are evident and supported by the mean values given in each panel.

a detour eye movement (error + correction) we can state, that the training clearly speeds the detour. The amount of time for the detour was on average 50 ms shorter after the training than before.

The errors occurring in the antisaccade task and the percentage of corrections following those errors are also improved by the training. The Fig. 13.7 shows the pairs of age curves. The error rate is reduced and the rate of correction is increased. Both variables indicate an improvement of the performance of the antisaccade task. We will consider the combined variable (pmis) later when estimating the success of the training.

Pre- versus Post-Training Values

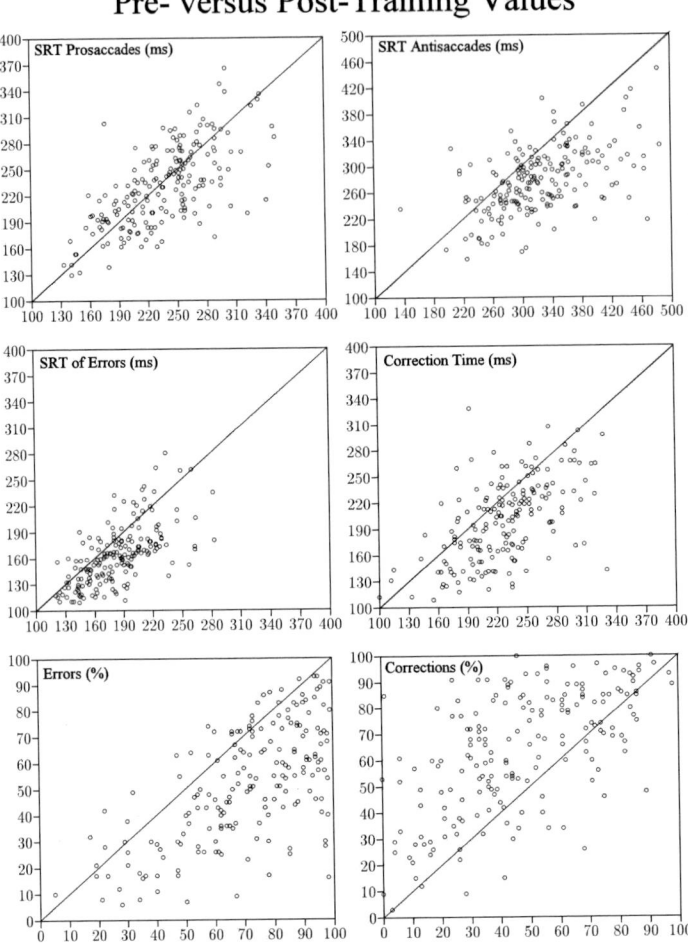

Figure 13.4. The 6 panels show the scatter plots of the 6 variables before an after the training. Data points on or close to the diagonal indicate "no change". SRT = saccadic reaction time.

13.4. Success of the Saccade Training

We have seen already that not all single variables of all subjects are systematically improved by the training. But it is not enough to know that the mean values show significant changes as we have seen in the previous section. What really counts is the percentage of subjects who were able to improve one or the other aspect of saccade control.

We know already that the reaction time of the prosaccades with overlap conditions are not systematically shortened or lengthened by the training. Therefore we first concentrate on the antisaccade task and treat the prosaccades from the overlap condition later.

To estimate the success rate we use the combined variable pmis, which describes the percentage of trials in which the subject failed to reach the opposite side of the stimulus by one or even by two saccades.

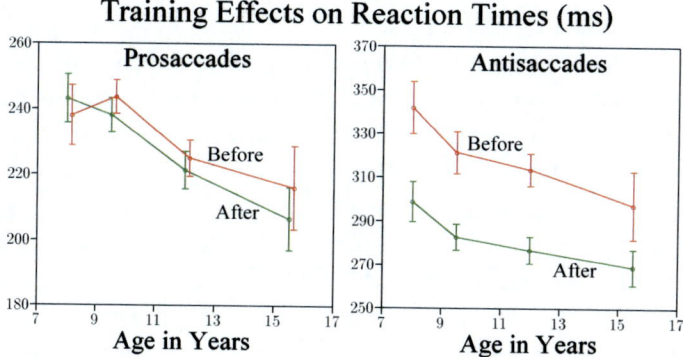

Figure 13.5. The figure shows the pre- and post- training values describing saccade control as a function of age. Little systematic training effects are seen in the reaction times of the prosaccades, clear shortenings of the reaction times are obtained for the antisaccade reaction times.

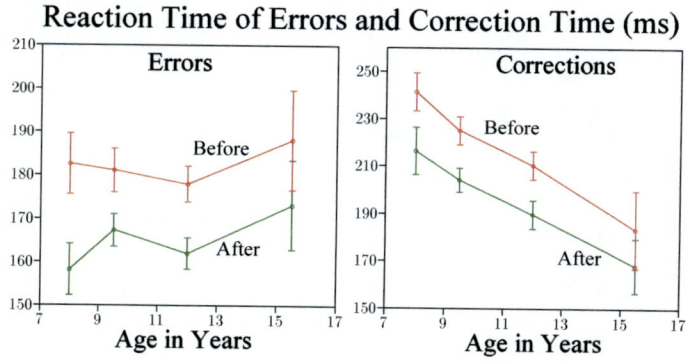

Figure 13.6. The figure shows the reaction time of the errors and the correction time as a function of age before and after the training. The amount of time training speeds up both the reaction and the correction time.

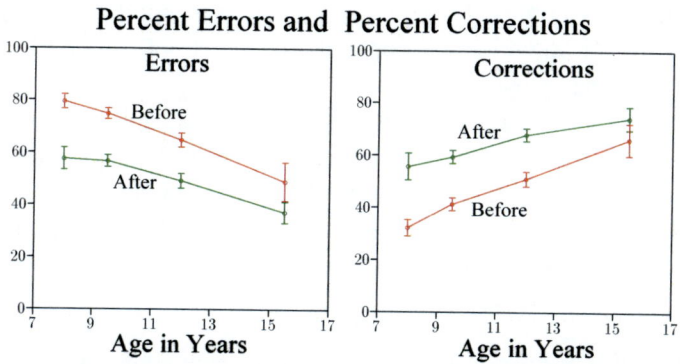

Figure 13.7. The figure shows the pairs of age curves of the error rate (left) and the correction rate (right) as a function of age before and after the training. The number of errors is greatly reduced and the number corrections after these errors is greatly increased.

The Fig. 13.8 show various aspects of measuring the success of the training in a quantitative way. First we show the scatter plot of the pre- and post- training values in the upper left panel. Almost all data points fall below the diagonal indicating that either the error rate was decreased or the correction rate was increased or both improvements were achieved. The pair of age curves are shown by the bottom left panel. It looks as though improvements were obtained in all age groups, but the differences between the two curves become smaller with increasing age. This may be caused by older subjects requiring longer training. The absolute value of the pre- (or the post-) training values and their scatter are also a functions of age. Therefore, we calculated the percent difference of the pre- and post-training values of the variable pmis. This is shown in the upper right panel. The age dependence is completely compensated by using the percent number of improvement. On average the improvement is 50% of the pretraining value.

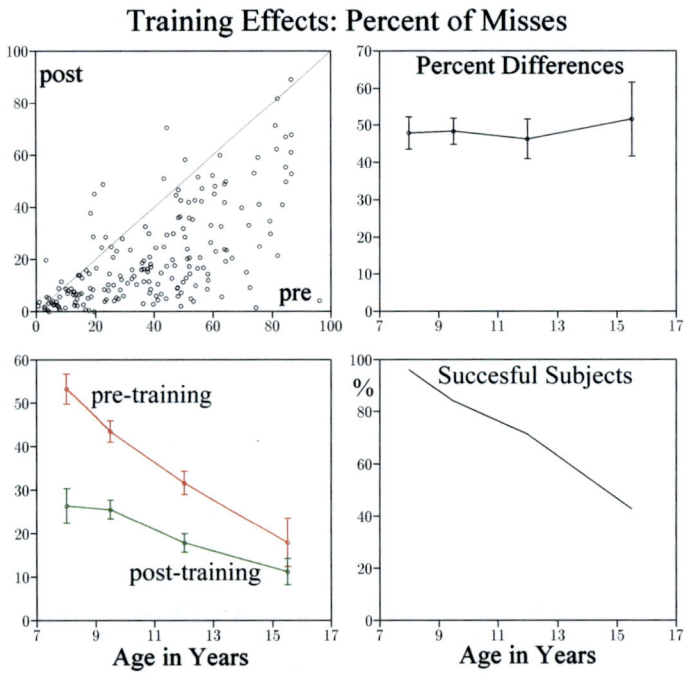

Figure 13.8. The figure illustrates the success of the antisaccade task training as measured by the percentage of trials, in which the opposites side was not reached. Post- versus pre-training values are depicted by the scatter plot in the upper left panel. Almost all data points fall below the diagonal line. The lower left panel shows the pre-and post-training age curves of the mean values. The differences between the curves depend on age. The age dependence is completely compensated by using the relative differences in percent as a measure (upper right diagram). The lower right diagram shows the percent number of subjects that have reached the range of the control subjects. This variable is the ultimate measure of success.

The ultimate measure of improvement is given by the percent number of subjects, who were able to reach the range of the control values. The curve in the bottom right panel shows that this percentage decreases steadily with age from very high values to about 50%.

In other words: almost all of the younger participants were able to reach the control range, while from the oldest group only 50% were able to reach the control range. The average success rate of all examined subjects was 77%.

Of course, one could even see in detail which of the antisaccade variables was improved to what extent. From this extended and detailed analysis we report only, that in less than 5% of the subjects no variable was improved at all.

13.5.   Effects of Training on Express Saccade Generation

As mentioned above, the training does not systematically change the variables describing the prosaccade task performance. Yet, single subjects may exhibit an extreme deviation of their pre-training values from the control values in one or the other direction. E.g. the reaction times of the prosaccades may be extremely short because of a preponderance of express saccades. From the scheme for elaboration of the training schedule we see, that in these cases practicing fixation would be the appropriate training. In the case of reaction times that are too long and/or extremely large scatter, a training of the saccade task would appropriate.

To see the corresponding effects on the performance of the prosaccade task we selected a small (N=7) group of subjects who practiced only the saccade task and no other tasks. Their reaction times were shortened by 62 ms and the percentage of express saccades was increased by 11%. In comparison, another group (N=77), who trained only the fixation and the antitask but not the saccade task, changed neither their percentage of express saccades nor their reaction time. Still another group trained the S- and the A-tasks, but not the F-task. Their reaction times and their express saccade numbers remained unchanged.

These examples show, that increasing the percentage of express saccades is achieved by using the saccade task. Such an increase is most of time not desired, one simply wants to consolidate and speed the reaction times of the prosaccades. And therefore, the S-task is rarely the only training.

A definite interest on increasing reaction times is needed in the cases of express saccade makers, whose only deficit is the preponderance of express saccades, but no difficulties with the antisaccades (they may make many errors with express reaction time, but they would correct almost all of them within short correction times). A reduction of the express saccades is possible by practicing the fixation task.

The Fig. 13.9 shows an example of a 10 year old boy, who exhibited 55% express saccades (express saccade maker). The mean value of the prosaccades to both sides were much too small (140 ms). Otherwise the diagnostic values were in the normal range. After an extensive training of fixation and a short period of the antitask, the reaction times of the prosaccades were slowed from 140 ms to 169 ms. One sees the large peak of express saccades before the training (left side, of the figure) which was reduced after the training. The peak is still clearly visible, but the number of express saccades is reduced to 40% and one clearly sees the 3 peaks after the training. In the antisaccade task the subject made 42% errors before and 21% after the training. The reaction times of the antisaccades were reduced from 307 ms to 236 ms. The subject also reduced the correction time after the errors to the left side from 212 ms to 152 ms. The percentages of corrections were high before and after the training.

This example demonstrates that the training can speed the saccadic reaction in one task and can slow them in the other task. These effects can be even be specific for the side of stimulation.

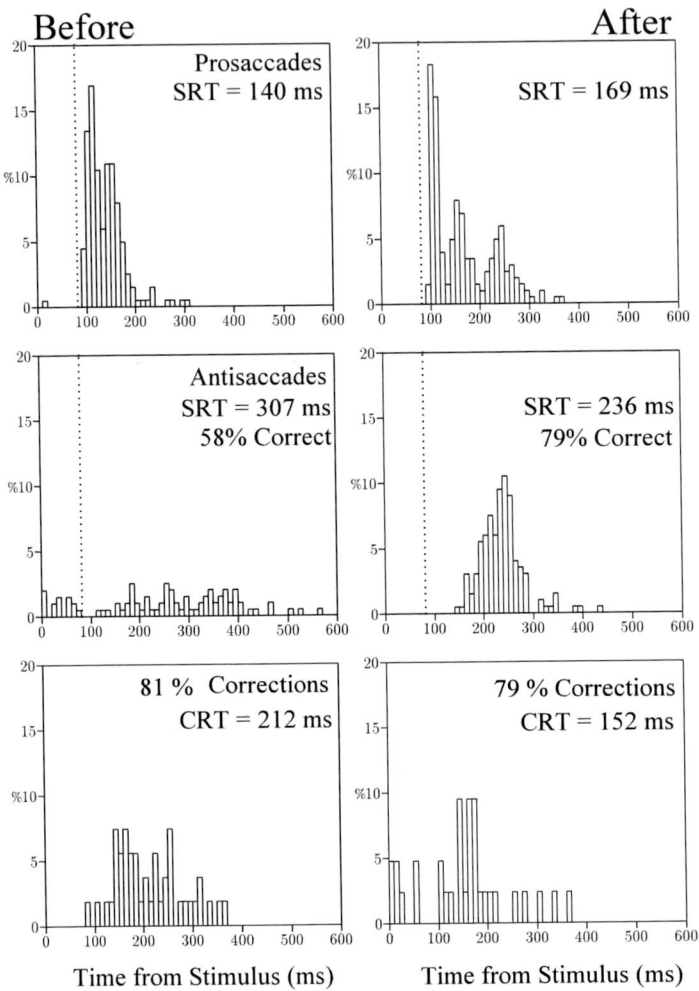

Figure 13.9. The effect of the training is shown by the distributions of the variables before (left) and after the training (right). Note that the reaction times of the prosaccades became longer, while the reaction times of the antisaccades and the correction times became shorte. Other details are described in the text.

13.6.   Training Effects on Fixation Stability

When we talk about the stability of fixation we have to keep in mind, that there are different ways of causing instability: one is a weak system of suppressing reflexive saccades. This aspect cannot be seen by analysing the eye movements during fixation. One needs to use a task, which requires saccades, in order to see, that these saccades have extremely short reaction times and contain high numbers of express saccades (indications for weak fixation), or to see that these saccades have normal or extremely long reaction times – indications for a fixation system that is too strong. These cases have been treated in the previous section.

The other types of fixation instability are the mono and the binocular instabilities. We consider the training effects on both types.

First, we look at pre/post training data of children, who completed their saccade training in a regular way under binocular viewing conditions. The differences between the pre- and post-training mean values did not reach significance, or for the mono instability nor for the binocular instability. The failure of positive training effects was even more evident, when looking at the percent difference of the pre- and post-training values.

Monocular Training

In a separate study the saccade training was changed for children with binocular instabilities. To stabilize the direction of gaze of both eyes to allow proper binocular fusion, we used the "old" well known method of occlusion. The fixation training was performed with one eye covered. If it was possible to know the weak eye, the weak eye was used for the training by covering the leading eye during the daily period of training (i.e. for about 10 min every day).

The Fig. 13.10 shows the data of a 13 year old girl before and after the monocular training. The right eye was the drifting eye and disrupted the binocular stability in 68% of the trials. The lower left diagram shows how the instability increased with the ongoing measurement indicating an effect of fatigue or of losing attention. The left eye was covered during the training. The right side of the figure shows the data after the training. Now only 14% of the trials disturbing binocular stability were left. An increase of the instability with increasing time was no longer evident.

The result of an analysis of the data from a group of 27 subjects, who received a monocular training is shown by Fig. 13.11. The diagram shows the pre- and post mean values of the binocular instability index. Clearly the stability has improved significantly. The inspection of the single data showed that all difference values were positive.

The data of the same subjects were also analyses with respect to their unwanted saccades (mono fixation instability), because it was possible that this aspect of fixation stability was also improved by the monocular training. The mean values before and after the training are shown by the Fig. 13.12. The difference between the columns is small and does not reach statistical significance.

We can look at the question, whether the monocular training improves also the monocular instability directly and in more detail. To make both variables comparable, the relative differences of the pre- and post-training values are calculated and expressed in percentage of the pre-training values. The Fig. 13.13 shows at the right side the scatter plot of these variables with the binocular values along the horizontal and the monocular values along the vertical axes. The correlation coefficient was -0.10 and failed significance by far (p=0.634). While only 6 data points fall above the diagonal (indicating more improvement of the monocular instability) 19 data points fall below the diagonal indicating less improve-

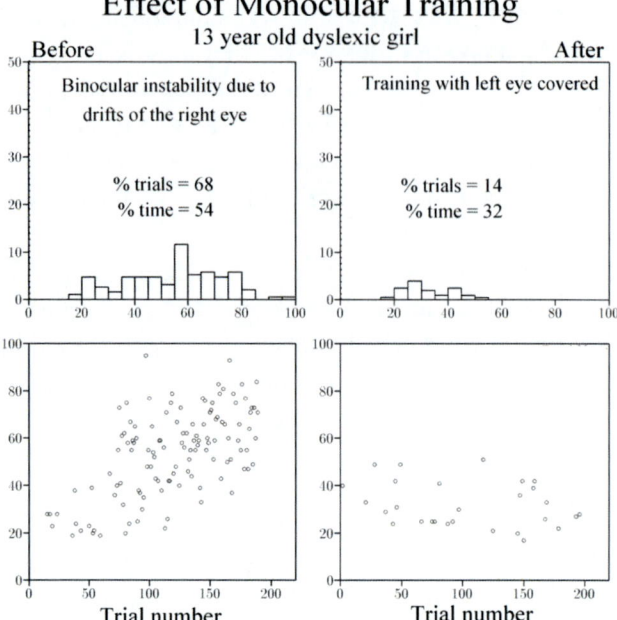

Figure 13.10. The figure shows the data of binocular instability of a 13 year old subject before and after the training with one eye covered. The upper panels show the distributions of the instability index before and after the training. Large numbers of trials exhibit instabilities (68%), the mean time of instability was about half of the analysis time window (54%). The lower left panel shows, that the instability increased during the time of the examination. The right panels show the results of the same analysis after the training.

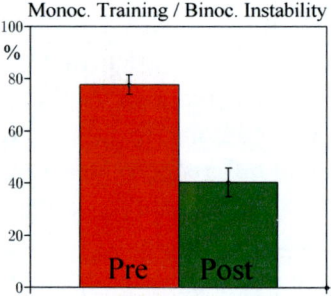

Figure 13.11. The figure illustrates the effects of a monocular training on the binocular stability of a group of 27 subjects. The index of binocular instability is decreased from almost 70% to less than 40%.

ment in the monocular instability than in the binocular instability. Of these, 2 fall at the horizontal line and 5 or 6 points fall below indicating that in these cases the monocular instability was even worse than before the training.

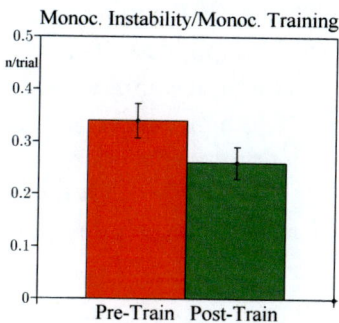

Figure 13.12. The pair of columns show the pre- and post-training mean values of the monocular instability of fixation. The difference is not significant, the mono instability remained unchanged.

The left side shows the mean values of the percent relative differences. While the monocular instability was improved by only 19%, the binocular instability was improved by 50%. Earlier attempts to improve reading in children with difficulties in fine stereopsis were successful by using the same strategy: cover one eye during reading [Stein and Fowler, 1985].

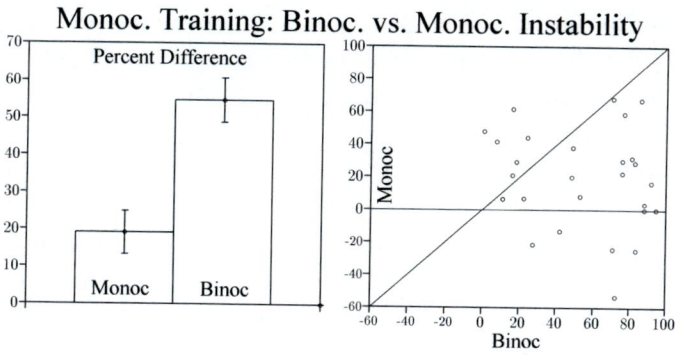

Figure 13.13. The left side shows the effects of the monocular training on the percent of improvement of the mono stability and the binocular stability. The right side shows the scatter plot of the variables. No correlation can be seen. See text.

From these considerations we can conclude again, that the two types of fixation instability are independent from each other and different in nature and that the binocular instability is systematically improved by a monocular training in almost all cases, while the monocular instability is improved by smaller amounts but not in all cases.

Finally, we want to know, whether a monocular training success in reducing the percentage of errors in the antisaccade task, would also reduce the monocular fixation instability. The answer is: there was no correlation between the variables (percent difference before and after the training). In about half of the cases there was no positive training effect in the monocular stability but a clear advantage in the reduction of errors. In the other half of the cases both variables exhibit a positive training effect

In conclusion: the monocular fixation instability must have a different origin as compared with the binocular instability and high percentages of errors must have still another reason. At the moment the best way of understanding this phenomenon is to assume that the intrusive saccades are "added" to the saccade generator in the brain stem when the central control of saccade generation and fixation stability cannot prevent these fast and small intrusive saccades.

13.7.   Specificity of the Saccade Training

A training of saccade control may have a general influence on the different components irrespective of the component that was part of the training. Even though this is not very likely, because we did not find systematic changes in the performance of the prosaccade task with overlap conditions and we know already, thet the generation of express saccades can be specifically influenced by training. Here we look for a more rigorous proof. We compare two groups of subjects: for one the saccade task but not the antitask were part of the training, while for the other group the antitask but not the saccade task was practiced. We compare the same variables for both groups: the pre-post difference of the reaction times obtained from the prosaccade task and the difference of errors made in the antisaccade task.

The Fig. 13.14 shows the results: when the saccade task was part of the training, the reaction times of prosaccades decreased by about 50 ms and the error rate did not change significantly. If the antitask was part of the training but not the saccade task, the opposite result was obtained: the prosaccades did not change their reaction times, while the error rate was reduced by about 20%.

Both groups of children went through the process of the diagnosis and they both used the training device and both had to focus their attention on the training every day for about the same amount of time. Yet the results are opposite. This indicates that there is no significant placebo effect in these data. Furthermore, this is support for the notion that the optomotor cycle indeed does consist of independent sub-functions probably provided by different structures in the brain. The effects of daily practice and the specificity of the training have been replicated in a recent independent study [Dyckman and McDowell, 2005].

These facts become important, when we look at the transfer of the saccade training to reading: the question of placebo effects must always be asked and we already have an answer to it from our study.

13.8.   The Training Protocol

Before the training one can not be sure that the children would improve their performance. Maybe they reach only moderate improvements or they do no improve at all. This may be because of different reasons. One is that they could not do any better, another could be that they did not follow the instructions. One has to keep in mind that most children performed the training with little or no control by their parents.

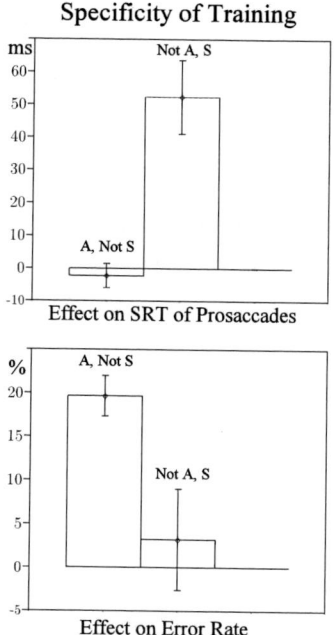

Specificity of Training

Figure 13.14. The two panels show the differences between reaction times of the prosaccades from before and after the training (upper panel) and the difference between the percentage of errors (lower panel). The left box in each panel shows the effect after a training of the saccade task but not the antitask (A, Not S) and the right box shows the effect after a training of the antitask and not the saccade task (Not, S). As the training schedules were opposite so were the effects of the training.

To see the result and to know how the training sessions were performed, the training instruments stored a complete protocol. The time, the day and the trained task, the level of difficulty in each task were stored together with the duration of the single sessions. In addition the response times were stored, even though the speed of the performing the task was not a criterion. But this makes it possible to see, when the children reacted too early not waiting until the end of the display sequence. This also usually showed the typical decrease in response time as the training proceeded.

All these data were used to elaborate a protocol describing the course of the training. Its success or its failure was described. In case of unsatisfactory results an extension of the training was recommended with the possible explanation of the failure.

The experience with the protocols has shown that a control of what the children really do during the period of the training was very important in finding the reasons for failures. Some children did not follow the instructions (poor compliance), misunderstood the instructions, did not achieve any substantial improvement, quit the training too early, or trained the wrong tasks. Therefore, in these cases it does not come as surprise, when they did not reach any progress in school. For all persons involved in the process of trying to help the child, it is important to know these details of the training.

The training protocol comes in graphical form, which is also described verbally in a written report, to make sure that the parents understand the result.

13.9.   Effects of Medication in Attention Deficit

Since ADS children are often treated by medication – e.g. with methylphenidate (Ritalin) – the effect oft this drug on saccade control was studied in a group of 10 -15 years old boys. This group was carefully diagnosed: the IQ as above 80, the EEG did not show any conspicuity of neurological diseases.

Their eye movements were recorded in the morning before they took their daily doses, i.e. more than 12 hours after the last intake. After the measurement they took their pill and after another 90-120 minutes the measurements were repeated.

To see the effect of the medication on saccade control the data of the sessions of all subjects were collapsed into one data file allowing to construct the distributions of the different variables [Klein et al. 2002]

The Fig. 13.15 shows the reaction times of the prosaccades with overlap conditions (left side) and of the antisaccades with gap conditions (right side). The distributions obtained without the medication are shown by the upper panels, those with the medication by the lower panels of the figure. One clearly sees the 3 peaks in the distribution of the reaction times of the prosaccades. The medication increases the first peak (representing the express saccades and reduces the long latency tail of values above 400 ms. This results in a shortening of the mean values of the all prosaccadic reaction times by 22 ms from 247 ms to 215 ms. The antisaccadic reactions times are also reduced by 43 ms from 317 ms to 274 ms. The whole distribution is consolidated.

The distribution of the reaction times of the errors and the distribution of the correction times are shown in Fig. 13.16. A shortening of the error reaction time by 30 ms reflects the corresponding shortening of the reaction times of the prosaccades in the overlap condition. With medication the long tail of values above 300 ms has disappeared almost completely. The errors occur faster but in smaller numbers: the error rate is reduced from 58 % to 43 %. The corrections times are also reduced by 33 ms from 204 ms to 171 ms and the percentage of the corrective saccades following the errors is increased by 20 % from 63 % to 83 %.

In conclusion from these comparison we see that the performance of both saccade tasks has improved. There is one exception, though: the increase of the percentage of express saccades is unwanted, because we know, that too many express saccades in the prosaccade task increase the chances of errors in the antisaccade task.

To quantify this differential effects of the medication we analysed the data by classifying the off-limit variables according to their functional significance:

- F-deficits suggest a training of fixation, because of a large number of express saccades.

- S-deficits suggest a training of the saccade task because of too long reaction times in the prosaccade task, especially because of too many saccades with extremely long reaction times and correspondingly large scatter.

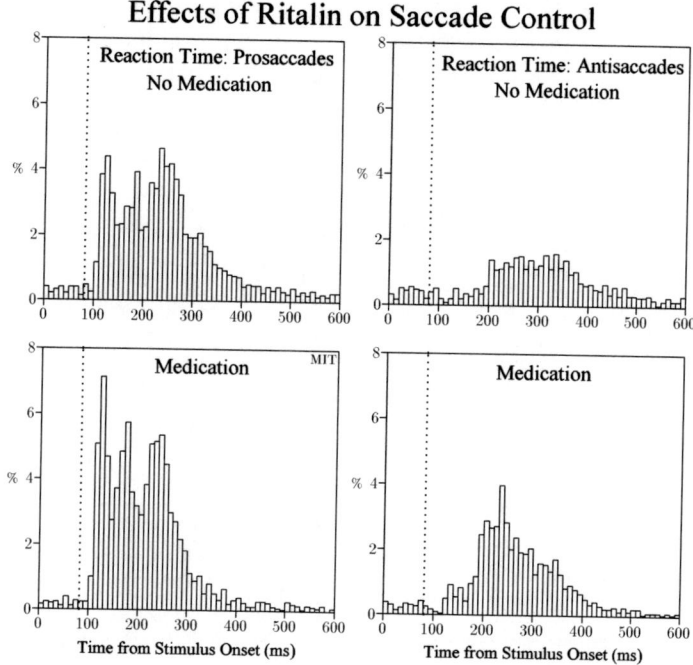

Figure 13.15. The figure shows the distribution of the reaction times of the prosaccades and the of the antisaccades of the whole ADS-group without and with the medication. Note, that the 3 peaks remain clearly separated and at about the same position of the x-axis, but their sizes change with the medication.

- A-deficits suggest a training of the antitask, because of too many errors and too few corrections of the errors, or because of reaction times and correction times that are too long.

The type of deficits were counted separately for all subjects with and without medication. The result is illustrated in Fig. 13.17. The number of F-deficits as indicated by the first pair of columns increased. The increase is not dramatic, but the tendency towards more express saccades as an effect of the medication is clearly visible. The S-deficits are decreased because the medication enables the children to generate their prosaccades within a narrow time window after stimulus onset thereby reducing the scatter and decreasing the mean values of the reaction time. The greatest effect is seen in the number of A-deficits. These are the variables describing the ability to generate saccades by conscious decision and against the direction of the reflex.

The fact that the medication has different effects on different components of saccade control can be interpreted as further support for the notion of one brain structure (presumably the parietal cortex) serving for active fixation (being weakened by the medication) and another brain structure (presumably the frontal cortex) serving for voluntary generation of saccades. This aspect of the medication would make it very important to find the correct doses, because one effect is wanted the other is unwanted.

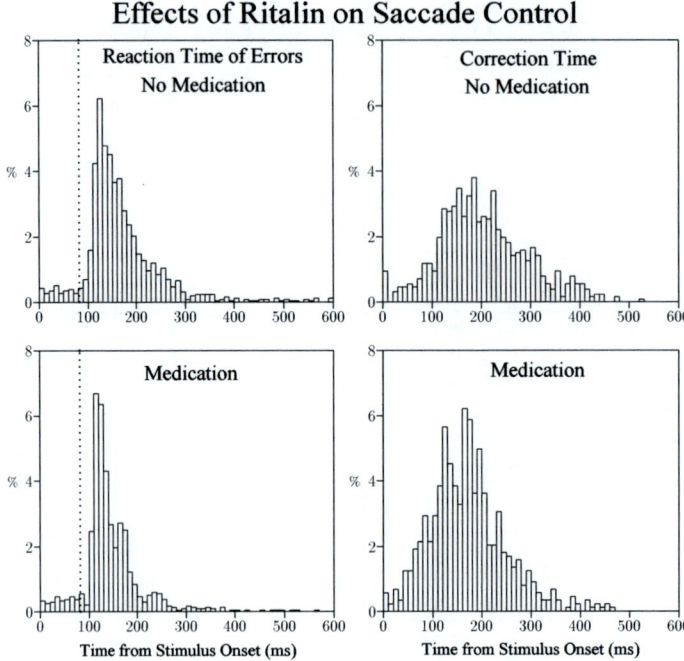

Figure 13.16. The figure shows the distributions of the reaction time of the errors and of the correction time with and without the medication.

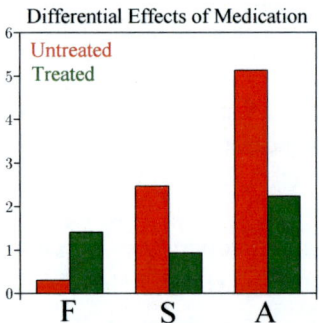

Figure 13.17. The columns show the differences between the different types of deficits in saccade control observed with and without medication. Note, that deficits in Fixation (F) are increased, while deficits in saccade (S) and antisaccade (A) control are reduced.

13.10. Training and Medication in Children with ADS

We already know, that an improvement of the system of saccade control can be achieved by daily practice. In ADS-children this may be difficult, because these children may not be able to concentrate on the training task every day for 10 minutes. In this case one might try to give the children the training when they are under the medication. In this case they may be able to do the training, but the effect of the training may disappear when the medication loses its positive effect after some hours. Here we review the results of attempts to answer these questions.

A group of ADS children (7 to 13 years old) was recruited. All members regularly used methylphenidate, all of them had deficits in saccade control and agreed to do the training required by their diagnostic values. The measurements of eye movements of all children were taken before the training without and with the medication as described in the previous section. The measurements were repeated after the training, again without and with the medication.

The Fig. 13.18 shows an example of data from a single subject (only the reaction times from the prosaccade task with overlap conditions are depicted). The pre-training data collected without medication is shown by the upper left diagram. Training created the upper right distribution, medication before the training created the lower left distribution. Here one sees the reduction of the reaction times by medication and the prolongation of the reaction times by training. The lower right panel shows the result of the measurement after the training and with medication. One learns from these observations and those from the other children and variables, that the training is possible for ADS-children and that it has a positive effect.

Since almost all effects described in this book depend on age, we also show the age curves of the data. This time we use the variable of the percentage of misses in the antisaccade task with gap conditions. This variable describes the performance of the antisaccade task by combining the percentage of errors and the percentage of corrections. Again the pre-training data without and with medication and the post-training data without and with medication are shown in Fig. 13.19. The upper left curve shows the data of the group before the training and without medication. The lower right curves shows the same curve and in addition the data after the training and with the medication. This gives the best results, i.e. the lowest values of the percentage of misses. The effect of the medication and the training effects add to each other.

In a further step the effects of the medication on the success of the training were investigated by having one group do the training under medication, the other group doing their daily training sessions before they took their daily doses, i.e. without a possible effect from the drug. Both groups were examined for their saccade control before and after their training and both were also examined without their daily doses and under the medication. Essentially there was no advantage of doing the training under medication. One of the reasons of course is, that the schedule of the training included also periods of doing the fixation task which prevents an increase of express saccades.

13.11. Saccade Training in Case of General Learning Deficits

The group of children suffering from general learning problems were also given the saccade training when necessary. The group contained only 19 children 9 to 15 years old. We show the variables of the prosaccade and the antisaccade tasks.

The fig. 13.20 shows the reaction times as a function of age. Despite the large scatter a reduction of the reaction times can be observed in both tasks.

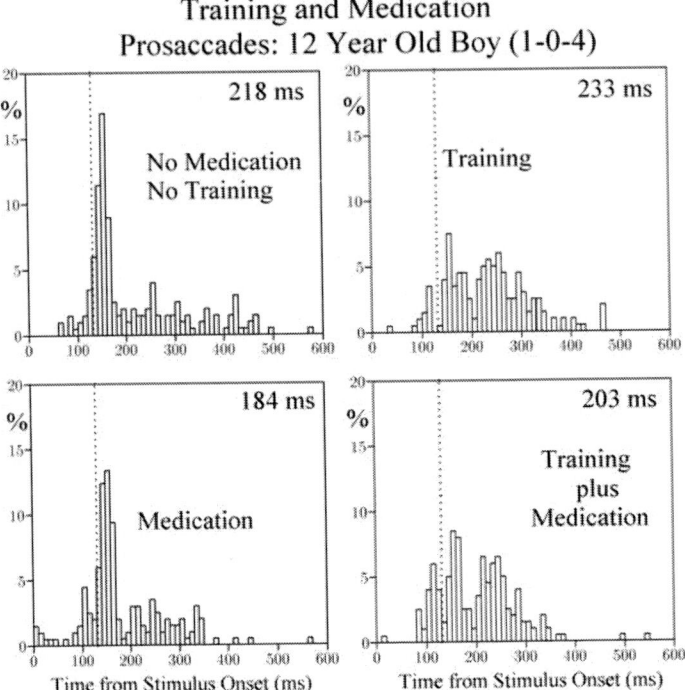

Figure 13.18. The distributions of prosaccade reaction times form a single subject are shown with and without training and with and without medication. The training schedule was: 1 week fixation, no saccade training, 4 weeks of antitraining. Medication and training alone have clear effects on the distributions. The strongest effect is obtained when medication and training are combined.

The reaction times of the errors and the correction times are shown in Fig. 13.21. The reaction times of the errors hardly changed at all, while the correction times decreased significantly only for the two younger groups.

Finally, we also look at the percentage of errors and the percentage of corrective saccades as shown in Fig. 13.22.. Both variables show clear improvements for the youngest group, but the oldest group could not profit from the training with respect to their errors and corrections.

To estimate the percentage of subjects, who were able to improve their saccade control, we use the variable of misses in the antisaccade task.

Out of the 19 participants 5 made more mistakes without correcting them after the training as compared to before the training. Four of them belong to the oldest age group. This means that the oldest group had the lowest training effect. The age curves also support this result, when looking at the other variables determined from the antisaccade task. Summarizing the data from all age groups 9/19 (47%) improved their antisaccade control.

In conclusion the effect of the training of saccade and fixation control was considerably smaller in this group of children with general learning problems as compared with the group of dyslexics or dyscalculics. The group also appears to be rather inhomogeneous

Effects of Training and Medication in ADS

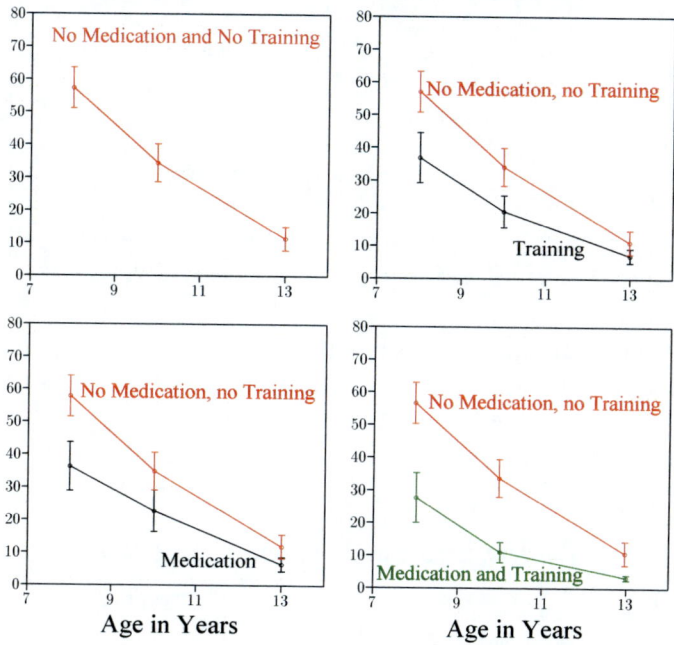

Figure 13.19. The figure shows the age curves of a group of ADS children. Upper left: No training and no medication. Upper right: after training, no medication. Lower left: no training, medication. Lower right: after training with medication. The original age curve from the upperleft panel is shown again in the other 3 panels to see the effect directly.

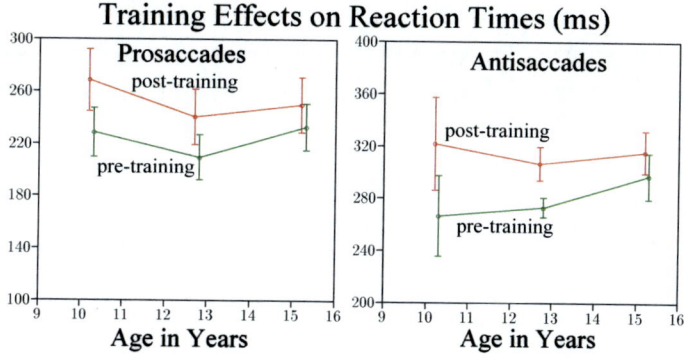

Figure 13.20. The reaction times of the pro- and antisaccades are shown before (upper curves) and after the training (lower curves) as a function of age in a group of children with general learning problems.

with respect to pedagogic aspects of their developmental state. Yet, one can always give a child a chance especially when they are younger than 10 years.

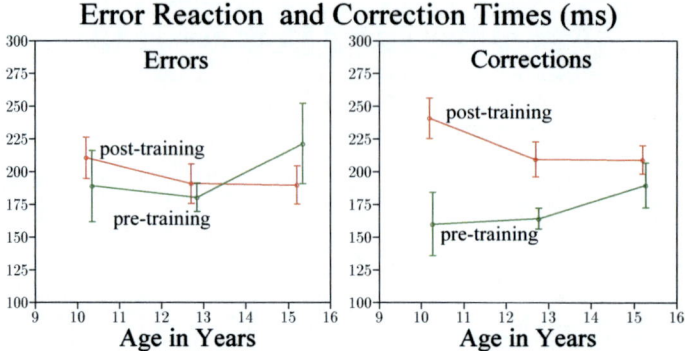

Figure 13.21. The pairs of age curves show the reaction times of the errors and the correction times before and after the training of a group of children with general learning
problems.

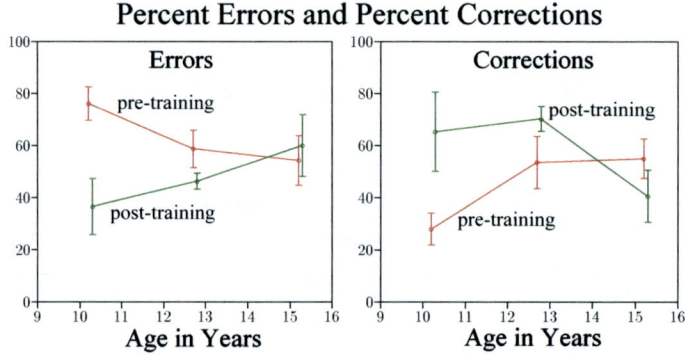

Figure 13.22. The percentage of errors and the percentage of corrections are shown by the
pair of age curves at the left and the right, respectively, in a group of children with general
learning problems.

PART IV

TRANSFER

Summary

The fact that the training effectively improves the perceptual and optomotor functions leaves the question open, whether there is also transfer to cognitive functions to be learned at school. This part describes the studies that have been conducted to answer these questions. Besides the consideration of the data of trained versus untrained groups, the data are also analysed person by person in a way much more similar to everyday life in school. The auditory, visual, and optomotor domains are treated separately.

The most interesting and most important part of this book is the question of transfer of the perceptual and optomotor training effects to achievements at school. First of all, this is not trivial. The aim of the training was an improvement of the special brain processing under consideration, which exhibited a deficit. Fluent reading for example needs more than correct saccade control. One can expect, however, that improvements in saccade control would make it easier to learn how to read. Regardless of what may be expected and what could not be expected, studies were conducted and the results are presented here.

One needs homogenous and age-matched groups of children, who received the same lessons of instructions at school or by private or other special teachers. One also needs parents, who are willing to have their child participating in a scientific study and who would accept, that their child may become a member of the placebo control or waiting group, instead of receiving the training immediately. Finally, one depends on whether or not the members from the beginning are still available at the end, when the measures of the pre-training tasks are taken again.

Therefore, studies of this kind often "suffer" from smaller numbers of subjects when compared with the size of the groups available for the question of development and diagnosis. In the following chapters we will describe several studies. Each of them concentrates on one combination of accomplishments at school and type of training:

- Spelling and auditory training

- Basic arithmetics and subitizing

- Reading and saccade training

These combinations were selected, because it appears that they are most closely related to each other. Yet, one has to keep in mind, that other combinations may play an important role as well and that in practice one must consider all aspects and one has to consider each individual child as a whole. For example, there may be general positive effects transferred from the training to learning at school, which are lost in the specific examination tests. E.g. one of the common achievements after the saccade training was an improvement of hand writing: the children for the first time could write along the lines, the letters were much more equal in size, one could read most words without doubts, which letters were used. This is not only an important accomplishment by itself, but it enables the child also to read what it had written and to check by itself for spelling or spelling errors. By this "detour" a child may improve spelling even though one would relate spelling more to auditory skills.

Chapter 14

Transfer of Auditory Training to Spelling

Summary

Improvements in auditory differentiation can be expected to transfer more or less directly to spelling. The results of a specific study are presented and it is shown, that indeed phonological awareness is improved and the spelling errors are significantly reduced in an experimental group as compared to a waiting and a placebo group.

The transfer of the auditory training to spelling was investigated by forming 3 groups: the experimental group (N=25), a waiting group (N=6), and a placebo group (N=10). The members of all groups had deficits in the auditory domains as described in the chapter on auditory development, but only the experimental and the placebo group were given training. The members of the experimental group trained those tasks that they failed during the examination, while the placebo group trained a visual fixation task. This way they had the same conditions in all details with the exception of the sensory system challenged by the training tasks. The details can be found in the published literature [Schäffler et al. 2004].

14.1. Phonological Awareness

Besides the spelling tests we also examined the phonological awareness using a standard task (HDLDT), which tests the differentiation of short rhyme words (only one syllable words) of the German language. E.g. wald – kalt, Stadt – matt, sind – Kind. Of course this is a test in German language and participants knowing the language will reach higher scores. All children in the study were native german speakers. By contrast, the low level auditory tasks, that were used for the training are completely free of any language.

The Fig. 14.1 shows the mean values of the pre- and post-training scores (expressed by their percentiles) of the phonological task (HDLDT) by the diagram at the left and the spelling tasks at the right for all three groups before and after the training. First of all, one sees that children who failed in the auditory tasks described here also failed the tests of phonological awareness: the mean value is p10 before the training. After the training they reached a mean value of p50.

This might not come as a surprise, because low level auditory discrimination can be regarded as a prerequisite for the differentiation of words. But it is important to notice, that deficits in phonological awareness may be caused by still "lower" (more fundamental) auditory skills, which are "below" language processing.

This view is supported by the fact, that the discrimination of one syllable rhyme words (HDLDT) is largely improved by the auditory training of non-language bound tasks. In fact, all children without any exception profit from the training when repeating the HDLDT after the training. The mean value of p50 by definition is the mean of the control group. However, all the post-training values were above the p16 limit, which indicates, that the transfer of training can be regarded as "complete".

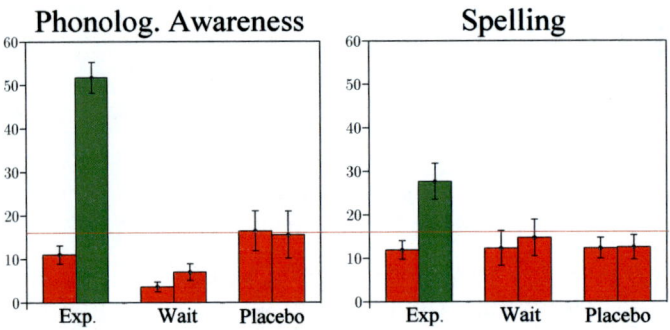

Figure 14.1. The figure shows shows the mean values of the pre- and post-training scores of the phonological task (HDLDT) and the spelling tasks for all three groups before and after the training. Note that only the experimental group reached values above p20.

14.2.   Spelling

The profit in spelling is also significant, but smaller when compared with the phonological value. To find the reason one has to take into account that the spelling task (DRT) contains 3 types of possible errors: type 1 errors are due to poor perception of the words, type 2 errors are due to lengthening vowels, and type 3 errors are due to poor knowledge of the German grammar rules.

The effect of the training on the spelling errors was analysed in detail by counting the 3 types of errors separately. The result is illustrated in Fig. 14.2. The green columns represent the data of the experimental group, the red columns those of the control group. The percent of error reduction is shown for the experimental group and the waiting and placebo groups, which are treated as one group for this purpose, because their results were not significantly different from each other, they both failed to profit from the training. Type 1 errors are significantly reduced in the experimental group, but the type 2 and type 3 errors did not show significant reductions.

Since the groups of this study are relatively small the rank order of the total group after the training shows the different effect on the groups. The percentage of spelling type1 error reduction is used as the variable for the rank order. The Fig. 14.3 shows the result.

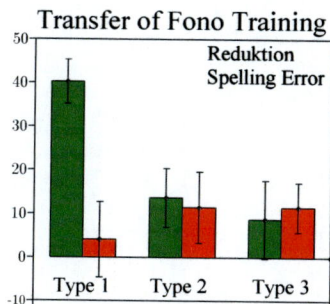

Figure 14.2. The 3 pairs of columns shows the reduction of spelling errors in percent separately for the 3 types of errors. The waiting and the placebo groups are combined, because they did not profit from the training. Note, that only the type 1 errors profit from the training.

The members of the experimental group (N=15) occupy the top 6 places of the rank order (N=30), while the last 11 places are occupied by the waiting or placebo group. Note also, that within these two groups the percentage of error reduction could be even negative signalling more errors after as compared with before the training. To estimate the success rate we use the criterion of 20: 75% of the experimental group, but only 20% of the two other groups reached this criterion. Even with a much stronger criterion of 50, half of the members of the experimental group but only 7% of the other two groups reached this criterion

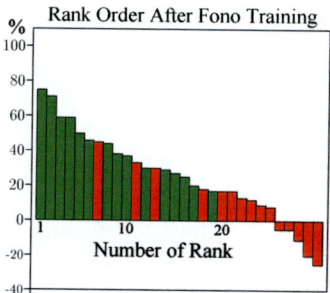

Figure 14.3. The figure shows the rank order of the percentage of reduced type 1 spelling errors (perception) after the training of the auditory tasks. From the experimental group 75% exceeded the criterion of 20, while only 20% from the other two groups reached this criterion. The heights of the columns is the percentage of error reduction. The green columns represent the members of the training group, the red columns represent the members of the waiting and placebo groups.

Chapter 15

Transfer of Visual Training to Basic Arithmetics

Summary

From the original hypothesis of the causes for dyscalculia it was argued, that an improvement in this visual domain might affect the achievements in basic arithmetic. This possible transfer of the visual training of subitizing was therefore studied in two groups of children with problems in basic arithmetic. The results are reported in this chapter and they support this view.

The transfer effect of training of subitizing was studied in a group of children with problems in basic arithmetic, but normal scores in reading and spelling. A group of 21 children in the age range from 7.5 to 9.0 years was recruited from a special school. All children suffered from deficits in subitizing and counting by memory. The group was divided in an experimental (N=10) and waiting group (N=11). The 2 groups were age matched with a mean age of 8.1 and 8.0 years, respectively. At the beginning all children took a standard test on basic arithmetic (DEMAT-II). This test was repeated after the training by using a second version of it to minimize effects of memory of the pretraining performance. (This caution was not really necessary, because it is the general experience, that children with problems in basic arithmetic do not profit from this kind of memory.)

Only the children in the experimental group received the training. All children went to school taking their ordinary lessons with no extra instructions on basic arithmetic. The Fig. 15.1 shows the differences of the scores reached in the DEMAT-II before and after the training. The first pair of columns illustrates the basic result of this study: while the experimental group gained 3 points, the waiting group gained nothing. If anything, the group lost points (statistically not significant). Thus the school lessons alone could not help the waiting group, the training plus school lessons improved the performance of the experimental group.

During the time period following the training, the experimental group kept visiting the school, while now the waiting group made the training of subitizing. The second pair of columns shows the differences in scores gained by the groups during this time period. Now, both groups gained points. The experimental group can continue to profit from the school lessons, the waiting group profit from the training and the school lessons.

After another period, while both groups continued to visit their school, the waiting group continued to gain points from school lessons, which had not been possible for them before the training (no data available from the experimental group). Finally, the total increase in scores is presented by the last pair of columns: both groups were about equally affected and gained about as many points during the whole study.

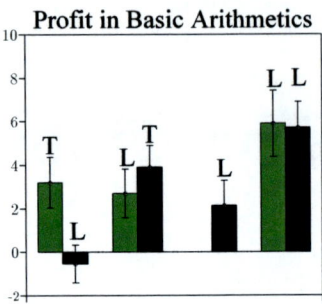

Figure 15.1. The effect of training of subitizing and number counting by memory is illustrated by the pairs of columns. The heights of the columns indicate the differences of points reached in the DEMAT-II test after the respective time period. The experimental group is shown in green, the waiting group in shown in black. For details see text.

We can also look at the rank order of the experimental group and the waiting group recombined in one as shown in Fig. 15.2. The first 6 places are occupied by members of the experimental group, the last 4 places are occupied by members of the waiting group. Note, that these 4 members lost points, i.e. their performance of the tasks decreased. If one uses a criterion of 3 points or more 60% of the experimental group and 9% of the waiting group reached this criterion after the first period of the study.

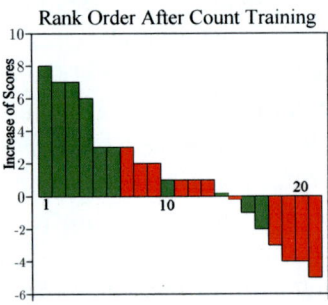

Figure 15.2. The figure illustrates the transfer effect of the count training on the profit of basic arithmetic skills by the rank order of the total group. The green columns represent the training group the red columns represent the waiting group.

More details of this work is available in the internet: www.optomlab.com

Chapter 16

Transfer of Saccade Training to Reading

Summary

A successful training of saccade control may be expected to have a more or less direct influence on reading skills, because the optomotor aspect of the reading process is improved and can be used more effectively than before the training. Indeed, a positive transfer was observed in a study of an experimental as compared to a waiting group.

40 children 7 to 12 years old suspected to have dyslexia were examined with respect to their reading and spelling capacities and with respect to their saccade control. Those who were classified as dyslexics and exhibited deficits in the frontal lobe component of their saccade control participated in the study.

The effect of training saccade control on reading was studied by dividing the group into an experimental group (N=11; age=9,8 y) and a waiting group(N=10; age=9,2y). The groups were almost but not quite age matched. The experimental group was given a training of saccade control, the waiting group had to wait to do their training after the study.

After the training both groups were recombined into one group receiving private les-sions in reading for another 6 weeks. The teacher did not know who was a member of which group.

After this period all measures were taken again from all members of both groups. In particular, the reading scores gained by counting the reading errors were analysed. The pre- and post-training values were compared between the two groups.

The Fig. 16.1 shows the result. The left side depicts the mean values of the differences between the pre- and post-training scores, the right side depicts the mean of the percentage of errors gained by the training.

In both panels, the first pair of columns considers the data of the complete groups ir-respective of their age. Clearly, the experimental group lost more errors than the waiting group. A more detailed analysis showed a significant age effect in the experimental group: older children profit much more from the training than younger children. A smaller age effect was obtained in the waiting group with the important difference that the sign of the correlation coefficient was opposite in the two groups. Therefore, in the second pair of columns the 7 year old children were eliminated from the analysis, in the third pair the 8 year olds were eliminated, and in the fourth pair the 9 year olds were eliminated. One sees that the effect of the training becomes larger. The percentage of error losses (right panel) gives a clearer picture than the differences of the errors, because calculating the percent values compensates already part of the age effects (all children made the same test, the test result was not normalized for age, the younger children made more errors than the older ones). The application of an analysis of variance (ANOVA) gave a significant age effect in the differences of the groups (p=0.01) and a significant difference between the groups.

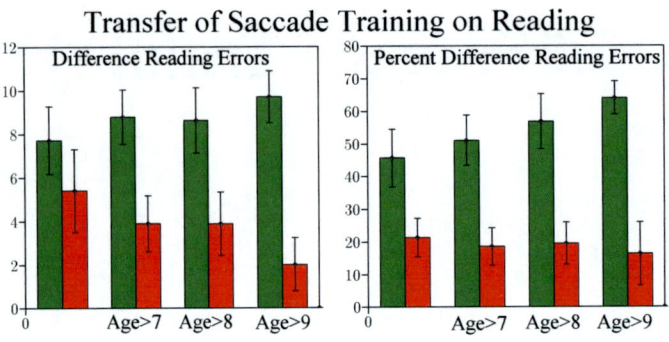

Figure 16.1. The figure shows the effect of the saccade training on the reading errors. The left side depicts the differences between the pre- and post-training errors, the right side shows the percent difference. The green columns represent the experimental group the red columns the waiting group.

As in the cases of the auditory training and the count training we look at the rank order of the members of the total group. The Fig. 16.2 shows that 7 members of the experimental group occupied the top positions. Almost 2 thirds of the experimental group reduced their errors by 50%, but not a single member of the waiting group reached this criterion.

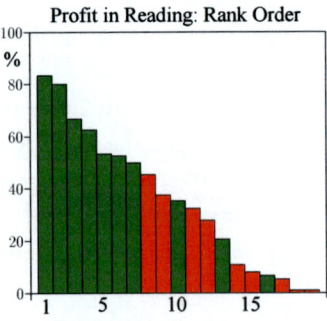

Figure 16.2. The figure illustrates the effect of the saccade training on reading by showing the rank order of the total group. The green columns represent the training group, the red columns represent the waiting group. Note: not a single member of the waiting group, but 64% of the experimental group reached 50% error reduction.

The original work is described in detail by a paper presented on our web page www.optomlab.com

In conclusion from these chapters we have seen that the 3 training procedures described in this book transfer to spelling (auditory training), to basic arithmetics (training of subitizing), and to reading (fixation and saccade training).

In practice, a given dyslexic child may often have deficits in more than one domain and therefore needs training in more domains. Of course even when the training in all domains is successful, we cannot expect a transfer rate of 100%, because the deficits in spelling, reading, or basic arithmetic may have still other causes, that could not be "repaired" by this training.

Conclusion

At the end of this book one might ask the question of what has been gained and what is left open for the future in the field of auditory, visual and optomotor functions and their relationship to problems in learning at school.

The thorough examination of perceptual and optomotor functions needed for learning at school has shown, that specific learning problems may be accompanied by deficits in fundamental processing of sensory information. These deficits are located in the brain, way below the level of language processing. However, the question of the causes of these problems is not completely answered. All we can conclude at the moment is, that improvements of the quality of these fundamental processing functions leads to facilitation of learning at school. For one or the other child this is the end of the struggling at all, in other cases the struggling might be at least reduced to an extent in which a normal school career is possible. From a pragmatic point of view, of course, this is the only thing that counts. A valid scientific explanation of the causes for the learning problem would not necessarily lead to a method of remediation. For example, the knowledge that genetic factors are causal in a special case of a dyslexic child does not help at all to solve the problem. But if one knows, that perceptual and/or optomotor control problems accompany the learning problem, even when caused by genetic factors, it might be possible to overcome these deficits and to facilitate learning how to read and write. The brain may develop strategies, which allow to reach the goal by "unusual ways".

The scientific research on learning and learning problems will continue, more methods for help will become available. Neuroscientists will find out more about learning of higher brain functions and cognitive skills. Maybe, one day we will be able to prevent the occurrence of learning problems altogether. In cases of deficits in the acquirement of language during the early years of life, one knows that the chances of the occurrence of learning problems at school are increased as compared with other children. Yet, the next step must be to find methods which allow to overcome the problem, hopefully before school begins.

By carefully looking at the data presented here some other questions in dealing with learning problems at school are solved. For example, the question: "are deficits in the auditory system responsible for dyslexia?". The question seems to require an answer of either "Yes" or "No". However, the question is already asked inadequately for several reasons. (i) the answer may be "yes" in one child and "no" in another. (ii) the answer may be "yes" for some components of the auditory processing and "no" for other components. (iii) the answer to the question may depend on age.

Another example: are children with dyscalculia impaired in subitizing? The answer may be "no" when one looks at the percentage of correct responses for 1 to 3 items, but it may be "yes" when looking at the response time for the same item numbers. The question already suggests that there is only one answer. But there are two aspects and correspondingly two variables, which – in this case – result in opposite answers.

Similarly, when considering saccade control, one arrives at different answers depending on what has been assessed: the reaction times of the prosaccades show no systematic differences, but the reaction times of the antisaccades are significantly longer in dyslexics as compared with age matched controls. The errors rate measured in the antisaccade task is clearly different in the two groups at an age above 8. For younger subjects there is no systematic difference, but single children may show extreme deviations by producing only prosaccades.

Another reason for discrepancies in the judgement of deficits in dyslexia or dyscalculia has to do with misinterpretations of the results of specific studies. If a group of dyslexics and a group of control subjects were assessed for their auditory skills and the comparison of the two groups exhibited "no significant difference", this does not necessarily mean that none of the dyslexics showed any deficit in this auditory test. It may be, that a certain fraction of the dyslexic group showed quite strong deviations from the mean value of the controls in one direction and another part of the dyslexic group performed much better than the controls. In this case the correct result of the study is: "a percentage of subjects among the dyslexics" exhibits deficits in the auditory domain.

Another example: If the efficiency of a medication is tested in an experimental and a control group, the result may be, that "a significant difference" was found. This does not imply, that the medication was effective for all subjects. It may even be that the difference between the mean values of the two groups was negligibly small for any practical purposes, but a statistical significance was still obtained because of the large number of participants in the study. In this case large numbers of the test and the control subjects reached the same values. For these the effect of the medication was not different at all.

Conclusions that are not really justified by the data lead to "artificial" contradictions. One way out is the inspection of the distributions of the values to be compared. In the parts "Deficits" and "Training" attempts were made to observe these cautions when presenting the data by calculating the percentage of subjects in a given group who failed a certain well defined criterion. The prize for the correctness of the data analysis is the complexity of the results and the need for detailed verbal descriptions. But after all: the complexity of the descriptions reflect the complexity of the brain functions involved in perception, optomotor coordination and learning.

The Appendix explains some of the basics statistics and shows that appropriated analysis and adequate description of the data, which help quite effectively to minimize misunderstandings, "artificial contradictions" and their consequences for subjects, who need proper diagnosis and effective therapy.

Appendix

Basic Statistics

This section explains a few terms which are frequently used in statistics. Invalid conclusions often rest on a lack of understanding, what is meant by those terms. For example, a "significant difference" between two sets of variables does not imply, that all single values of one set are different from all single values of the other set. There are two possibilities how a difference can become "statistically significant": the difference can be large or the number of measurements can be large. In practice: in a case of a therapeutic method which leads to a significant improvement of a variable under consideration does usually not mean that every single subjects using this method profits from it. In the opposite case, where a therapeutic method does not lead to a significant improvement, this does not imply, that no subjects profit.

Characteristic of Groups

In many cases, each member of a specified group can be assigned a single value, e.g. the error rate of the antisaccade task. To characterize the group, one looks for a few numbers that carry some knowledge of the whole set of data. The first step should always include the inspection of the distribution of the data.

The Distribution

The variables will always scatter within a range between a smallest and a greatest value. Not all values will appear with the same frequency. Most values will concentrate somewhere in the middle of the range, while towards the lower and the upper limit fewer values will be found. To visualize the frequency distribution, one divides the total range in small sections and counts the number of values falling within each of these sections. Such a section is called a bin. It must be small enough to allow a differentiation between the smaller and the larger values, and it must be large enough to catch a meaningful (minimum) number of values. To get a graphic picture of the distribution one draws a horizontal axis (x-axis) for the range of the variable and above each bin range a box with a height as specified by the absolute or relative number of values within the respective bin. The Fig. A.1 shows examples of two distributions.

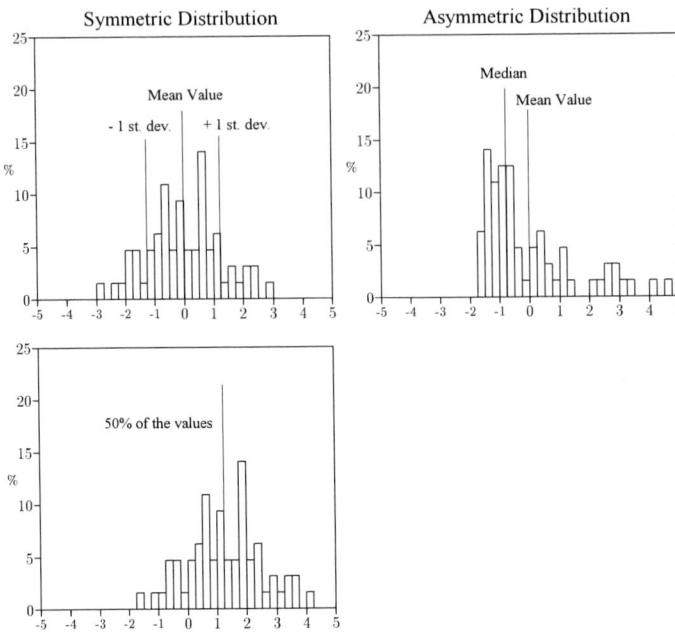

Figure A.1. The panels show distributions of single values generated artificially. The upper left panel depicts a symmetric distribution close to a Gaussian form. Its mean value and the standard deviation are shown. Below one sees another distribution. It is identical the same as the upper distribution, but the mean value is increased from zero to 1.25, i.e. by one standard deviation. The right panel shows an asymmetric distribution and the position of the mean value and the median. 50% of the values are located to the left of the upper mean value plus one standard deviation. For details see text.

The Mean Value

In cases, where the distribution is about symmetric around the middle of the total range (ideally a Gaussian distribution, or a "normal" distribution) one can calculate the arithmetic mean value. It is defined as the sum of all single values divided by the number of values. The upper left panel of Fig. A.1 shows such a distribution. Its mean value is zero as indicated by the vertical line.

The Standard Deviation

Most of the time the mean value is not enough to characterize the group. One needs in addition a measure of the scatter of the single values. The most commonly used measure is the standard deviation. It is calculated by a mathematical formula. If the distribution has a gaussian form two thirds of the values are found between minus and plus one standard deviation from the mean value and 95% within two standard deviations. The Fig. A.1 indicates the standard deviation by the vertical lines to the left and right of the mean value.

One can always calculate the mean value and the standard deviation, but there are cases in which these measures do not carry an adequate description of the group. This happens for example when the distribution is not symmetric but skewed to one or another side. In such cases other measures are needed to characterize the group.

The Standard Error

While the standard deviation specifies a range in which another measurement value will probably fall, the standard error interval specifies the range in which the mean will probably fall if all measurements of the whole group are repeated with another group. It also gives an estimation of how close the calculated mean value of a small subgroup is to the "real" mean value that one would get when measuring the whole group. If one divides the standard deviation by the square root of the number of values one arrives at a number, which is called the standard error. It gives the range within which the "real" mean values will be found with a probability of 68%. This definition immediately shows that the standard error becomes smaller with increasing numbers of values and eventually converges to zero. The standard deviation, on the other hand, will converge to a finite value.

The Median and the Rank Order

The median is another measure of a group of single values. It is based on counting. It follows the same rules as one knows from the rank order in sport events. What counts, is who is first and second, or who is last in a run. The time they needed is only used to construct this rank order. As one normalizes the total number of subjects (values) to 100% one arrives at a rank order with 0 as the lowest and 100 as the highest rank. The middle is 50. The value which corresponds to 50 is called the median. In a normalized rank order the values are often called the percentile.

The rank order and the median are used in cases of asymmetric distributions. An example is shown in the right panel of Fig. A.1. Its mean value is zero, but its median is -1.0 as indicated by the vertical lines.

The scatter of the values in an asymmetric distribution can be defined by specifying rank orders above and below the median. In case of a Gaussian distribution (and any symmetrical distribution) the median is identical with the mean value. One standard deviation below (above) the mean value corresponds to a rank order (percentile) of 16 (84). Using the rank order and the percentiles has the advantage, that the result is independent of the form of the distribution.

Characteristic of Single Subjects

In scientific studies the group statistics are important, because they are considered to allow to judge, whether a certain effect is specifically related to a voluntarily implemented action or must be regarded as being reached by chance because random deviations just happened to accumulate in one direction. In a medical setting one wants to classify the situation of a single subject. It is of no interest for the patient to know, that she/he is a member of a group, which differs statistically from another group. She/he wants to know where she/he is located in a group of control subjects. She/he wants to know, what the chances are that she/he will find effective help.

Deviation from the Mean Value

If one knows the mean value and the standard deviation one can classify a single subject by the distance of the single value given in units of the size of the standard deviation. On a horizontal scale, the null position in the middle indicates the mean value. Equal units right and left indicate standard deviations. A single value within one unit left and right means that the value is "normal".

The Percentile

If the mean value and the standard deviation are inadequate one can use the median and the percentile to characterize the value of a single subject. One has to find the rank of the individual value within the rank order of the control group. This rank expressed in percent is called the percentile. By convention, high percentiles indicate high quality, low percentiles indicate low quality. In cases where the distribution of the control values is rather narrow, the percentile of an individual is very sensitive to small changes in the performance of the task. Whatever method one uses to classify a single subject, there are regions of transition of the measured value from "normal" to "conspicuous".

Comparison of Groups

Often one wants to compare two groups. After one has characterized the groups by mean values and standard deviations, the question arises, whether the difference between the mean values is significant or has been reached by chance. Obviously, the answer to this question depends not only on the size of the difference, but also on the scatter and the numbers of the values in each group.

Significance

To arrive at a quantitative measure, the significance p has been defined. It is a number between 0 and 100%. It tells us the probability by which the hypothesis that the two groups are the same is wrong. Small values of p (below 5% or below 1%) indicate that the two groups are statistically different, because the probability that they are the same is small.

With the help of the Fig. A.1 we can see, what it means and what it does not mean, when two distributions are "significantly" different. The mean values of the upper and the lower distribution are different by one standard deviation. In case the shift of the mean value was due to a medication, one would like to argue, that the medication was effective. Yet, 50% of the values of the lower distribution are still within one standard deviation of the pre-medication values. The "real" result therefore is: the medication was effective for 50% of the patients.

The question, whether or not a certain group of subjects (e.g. dyslexics) suffers from a certain deficit (e.g. in the auditory domain) must receive a similarly detailed answer: the group of dyslexics differs significantly from the control group, but only 50% of the test group performed below the "normal" limit (by more than 1 standard deviation, or: below the p16 limit) of the controls.

Analysis of Variance and Factor Analysis

More sophisticated methods for analysing group differences are also available. The analysis of variance (ANOVA) tries to estimate how much of the variance of the values of the total group is due to fact, that the subjects belong to one or the other group. Unknown factors influencing the values may have caused the variance such that the membership of the groups do not explain the variance or only a small part of it. The factor analysis tries to find, how many independent factors are needed to explain the total variance.

These methods have been used in the analysis of the data published in international journals. In the data presented in this book besides the membership of the groups, age is the most important independent variable. By looking at the pairs of age curves the reader, has a pretty good idea of the significance of group and age. Since in most cases the number of subjects under consideration is pretty large (100 or more), one does not need to worry about significances. In many graphs, not only the mean values, but also the standard errors are shown (by vertical bars). If two values differ more than approximately twice the size of their standard error, then the difference can be expected to be significant.

Correlations

In certain sets of data one variable may be related to another, for example many variables discussed in this book are related to age. To see such possible relations one looks at the scatter plot of the two variables. Fig. A.2 shows a special example, in which the statistic calculations alone will lead to an invalid conclusion.

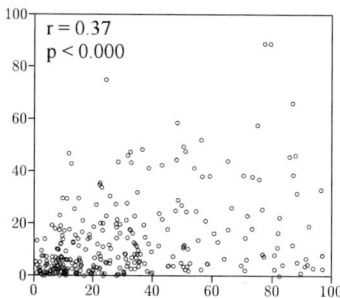

Figure A.2. The figure shows a scatter plot of two variables. Most data points fall below the diagonal. While the correlation coefficient is highly significant, inspection of the data show that only large y-values are related to large x-values, but not vice versa. See text.

Each dot represents the values of two variables from one person. The horizontal position represents the value of the variable of the x-axis, the vertical position represents the value of the variable of the y-axis. To characterize the relation one calculates the correlation coefficient r and its significance p. The correlation coefficient r is a measure for the strength of the correlation, its value is in the range -1 to +1. Values near 0 mean a weak correlation. The significance p is a measure to estimate, whether the correlation may have been obtained by chance, its value is in the range 0 to 1, values near 0 mean high significance. In the case of the figure the correlation coefficient is 0.37 and $p < 0.0001$. The reading of these two numbers is: the variables exhibit highly significant correlations with the implication, that if variable x is large so is variable y and vice versa. However, by inspection of the figure one sees immediately that this statement does not adequately describe the data. The correct description is: if variable x is large, variable y may take any number between 0 and 100. But if variable y is large, then variable x is also large. In other words: the correlation exists only in one direction. Unless one looks at the scatter plot, this result will be missed and misinterpretations of the data or contradictions are pre-programmed.

List of Instruments

The following instruments have been used to collect the data presented in this book.

Diagnostic Instruments:

- FonoFix measures low level auditory functions

- FixTest measures dynamic vision

- CountFix measures subitizing

- ExpressEye measures eye movement control

Training Instruments:

- FonoTrain trains low level auditory functions

- CounTrain trains subitizing

- FixTrain trains dynamic vision and eye movement control

References

Biscaldi, M; Fischer, B; Aiple, F. (1994). Saccadic eye movements of dyslexic and normal reading children. *Perception* **23**: 45-64

Biscaldi, M; Fischer, B; Hartnegg, K. (2000). Voluntary saccade control in dyslexia. *Perception* **29**: 509-521

Biscaldi, M; Fischer, B; Stuhr, V. (1996). Human express-saccade makers are impaired at suppressing visually-evoked saccades. *J Neurophysiol* **76**: 199-214

Breitmeyer, BG; Ganz, L. (1976). Implications of sustained and transient channels for theories of visual pattern masking, saccadic suppression, and information processing. *Psychological Review* **83**: 1-36

Dehaene S (1997) *The number sense - How the mind creates mathematics*.

Dehaene S (1999) *Der Zahlensinn*. Birkhäuser Verlag, Basel,

Dehaene, S; Cohen, L. (1995). Towards an anatomical and functional model of number processing. *Math Cogn* **1**: 83-120

Dehaene, S; Dehaene-Lambertz, G; Cohen, L. (1998). Abstract representation of numbers in the animal and human brain. *Trends Neurosci.* **21**: 355-361

Dehaene, S; Piazza, M; Pinel, P; Cohen, L. (2003). Three parietal circuits for number processing. *Cogn Neurophysiol*

Dyckman, KA; McDowell, JE. (2005). Behavioural plastcity of antisaccade performance following daily practice. *Exp-Brain-Res* **162**: 63-69

Eden, FG; VanMeter, JW; Rumsey, JM; Maisog, JM; Woods, RP; Zeffiro, TA. (1998). Abnormal processing of visual motion in dyslexia revealed by functional brain imaging. *Letters to Nature* **382**: 66-69

Everling, S; Fischer, B. (1998). The antisaccade: a review of basic research and clinical studies. *Neuropsychologia* **36**: 885-899

Fischer, B. (1987). The preparation of visually guided saccades. *Rev Physiol Biochem Pharmacol* **106**: 1-35

Fischer B (2005) Subitizing and Counting by Visual Memory in Dyslexia. In: Nova Publisher (ed) *Dyslexia: Recent Research.*

Fischer, B; Biscaldi, M; Gezeck, S. (1997). On the development of voluntary and reflexive components in human saccade generation. *Brain-Res* **754**: 285-297

Fischer, B; Biscaldi, M; Otto, P. (1993). Saccadic eye movements of dyslexic adult subjects. *Neuropsychologia* **31**: 887-906

Fischer, B; Boch, R. (1983). Saccadic eye movements after extremely short reaction times in the monkey. *Brain-Res* **260**: 21-26

Fischer, B; Breitmeyer, B. (1987). Mechanisms of visual attention revealed by saccadic eye movements. *Neuropsychologia* **25**: 73-83

Fischer, B; daPos, O; Stürzel, F. (2003). Illusory illusions: The significance of fixation on the perception of geometrical illusions. *Perception* **32**: 1001-1008

Fischer, B; Gezeck, S; Hartnegg, K. (1997). The analysis of saccadic eye movements from gap and overlap paradigms. *Brain Research Protocols* **2**: 47-52

Fischer, B; Gezeck, S; Hartnegg, K. (2000). On the production and correction of involuntary prosaccades in a gap antisaccade task. *Vision Res* **40**: 2211-2217

Fischer, B; Gezeck, S; Huber, W. (1995). The three-loop-model: A neural network for the generation of saccadic reaction times. *Biol Cybern* **72**: 185-196

Fischer, B; Hartnegg, K. (2000). Effects of visual training on saccade control in dyslexia. *Perception* **29**: 531-542

Fischer, B; Hartnegg, K. (2000). Stability of gaze control in dyslexia. *Strabismus* **8**: 119-122

Fischer, B; Hartnegg, K. (2002). Age effects in dynamic vision based on orientation identification. *Exp-Brain-Res* **143**: 120-125

Fischer, B; Hartnegg, K. (2004). On the development of low-level auditory discrimination and deficits in dyslexia. *Dyslexia* **10**: 105-118

Fischer, B; Hartnegg, K; Mokler, A. (2000). Dynamic visual perception of dyslexic children. *Perception* **29**: 523-530

Fischer, B; Ramsperger, E. (1984). Human express saccades: extremely short reaction times of goal directed eye movements. *Exp-Brain-Res* **57**: 191-195

Fischer, B; Ramsperger, E. (1986). Human express saccades: effects of randomization and daily practice. *Exp-Brain-Res* **64**: 569-578

Fischer, B; Weber, H. (1990). Saccadic reaction times of dyslexic and age-matched normal subjects. *Perception* **19**: 805-818

Fischer, B; Weber, H. (1992). Characteristics of "anti" saccades in man. *Exp-Brain-Res* **89**: 415-424

Fischer, B; Weber, H. (1993). Express Saccades and Visual Attention. *Behavioral and Brain Sciences* **16**,3: 553-567

Fischer, B; Weber, H; Biscaldi, M; Aiple, F; Otto, P; Stuhr, V. (1993). Separate populations of visually guided saccades in humans: reaction times and amplitudes. *Exp-Brain-Res* **92**: 528-541

Fuster JM (1991) *The prefrontal cortex: anatomy, physiology, and neurophysiology of the frontal lobe*. Raven Press, New York,

Gezeck, S; Fischer, B; Timmer, J. (1997). Saccadic reaction times: a statistical analysis of multimodal distributions. *Vision Res* **37**: 2119-2131

Goldberg, ME; Colby, CL. (1992). Oculomotor control and spatial processing. *Curr Opin Neurobiol* **2**: 198-202

Grigorenko, EL; Wood, FB; Meyer, MS; Hart, LA; Speed, WC; Shuster, A; Pauls, DL. (1997). Susceptibilty loci for distinct components of developmental dyslexia on chromosome 6 and 15. *American Journal of Human genetics* **60**: 27-39

Guitton, D; Buchtel, HA; Douglas, RM. (1985). Frontal lobe lesions in man cause difficulties in suppressing reflexive glances and in generating goal-directed saccades. *Exp-Brain-Res* **58**: 455-472

Hallet, PE. (1978). Primary and secondary saccades to goals defined by instructions. *Vision Res* **18**: 1279-1296

Hallett, P. (1978). Primary and secondary saccades to goals defined by instructions. *Vision Res* **18**: 1279-1296

Hartnegg, K; Fischer, B. (2002). A turn-key transportable eye-tracking instrument for clinical assessment . *Behavior, Research Methods, Instruments, & Computers* **34**: 625-629

Hebb D (1949) The oranization of behaviour: A neuropsychological theory. Wiley, New York,

Klein, C; Fischer, B. (2005). Instrumental and test-retest reliability of saccadic measures. *Biological Psychology* **68**: 201-213

Klein, C; Fischer Jr, B; Fischer, B; Hartnegg, K. (2002). Effects of methylphenidate on saccadic responses in patients with ADHD. *Exp Brain Res* **145**: 121-125

Mayfrank, L; Mobashery, M; Kimmig, H; Fischer, B. (1986). The role of fixation and visual attention in the occurrence of express saccades in man. *Eur Arch Psychiatry Neurol Sci* **235**: 269-275

Mc Anally, KI; Stein, JF. (1996). Auditory temporal coding in dyslexia. *Proc R Soc Lond B Biol Sci* **263**: 961-965

Mishkin, M; Ungerleider, LG; Macko, KA. (1983). Object vision and spatial vision: two cortical pathways. *Trends Neurosci.* **6**: 414-417

Mokler, A; Fischer, B. (1999). The recognition and correction of involuntary saccades in an antisaccade task. *Exp Brain Res* **125**: 511-516

Munoz, DP; Everling, S. (2004). Look away: the anti-saccade task and the voluntary control of eye movement. *Nature Reviews/ Neuroscience* **5**: 218-228

Munoz, DP; Wurtz, RH. (1992). Role of the rostral superior colliculus in active visual fixation and execution of express saccades. *J-Neurophysiol* **67**: 1000-1002

Munoz, DP; Wurtz, RH. (1993). Fixation cells in monkey superior colliculus. I. Characteristics of cell discharge. *J Neurophysiol* **70**: 559-575

O'Regan, JK. (1990). Eye movements and reading. [Review]. *Reviews of Oculomotor Research* **4**: 395-453

Olson RK, Conners FC, Rack JP (1991) Eye movements in dyslexic and normal readers. In: Stein JF (ed) Vision and Visual Dysfunction, Vol 13, *Vision and Vision Dyslexia.* Macmillan, London, pp 243-250

Olson, RK; Kliegl, R; Davidson, BJ. (1983). Dyslexic and normal readers' eye movements. *J Exp Psychol [Hum-Percept]* **9**: 816-825

Pavlidis, GT. (1985). Erratic eye movements and dyslexia: factors determining their relationship. *Percept Mot Skills* **60**: 319-322

Pavlidis, GT. (1985). Eye movements in dyslexia: their diagnostic significance. *J Learn Disabil* **18**: 42-50

Poggio, GF; Fischer, B. (1977). Binocular interaction and depth sensitivity in striate and prestriate cortex of behaving rhesus monkey. *J-Neurophysiol* **40**: 1392-1405

Rayner, K. (1978). Eye movements in reading and information processing. *Psychol.Bull.* **85**: 618-660

Rayner, K. (1985). Do faulty eye movements cause dyslexia? *Develop Neuropsychol* **1**: 3-15

Rayner, K; Murphy, LA; Henderson, JL; Pollatsek, A. (1989). Selective attentional dyslexia. *Cognitive Neuropsychol* **6**: 357-378

Rayner, K; Slowiaczek, ML; Clifton C, Jr; Bertera, JH. (1983). Latency of sequential eye movements: implications for reading. *J Exp Psychol [Hum-Percept]* **9**: 912-922

Reichle, ED; Rayner, K; Pollatsek, A. (2003). The E-Z-Reader model of eye-movement control in reading: comparison to other models. *Behavioral and Brain Sciences* **26**: 445-526

Saslow, MG. (1967). Latency for saccadic eye movement. *J Opt Soc Am* **57**: 1030-1033

Schäffler, T; Sonntag, J; Fischer, B. (2004). The effect of daily practice on low-level auditory discrimination, phonological skills, and spelling in dyslexia. *Dyslexia* **10**: 119-130

Schiller, PH; Sandell, JH; Maunsell, JH. (1987). The effect of frontal eye field and superior colliculus lesions on saccadic latencies in the rhesus monkey. *J Neurophysiol* **57**: 1033-1049

Simon, TJ; Peterson, S; Patel, G; Sathian, K. (1998). Do the magnocellular and parvocellular visual pathways contribute differentially to subitizing and counting? *Percept Psychophys* **60**: 451-464

Starkey, P; Cooper, RG. (1980). Perception of number by human infants. *Science* **210**: 1033-1035

Stein, J. (1993). Dyslexia–impaired temporal information processing? *Ann N Y Acad Sci* **682**: 83-86

Stein, J; Fowler, S. (1985). Effect of monocular occlusion on visuomotor perception and reading in dyslexic children. *LANCET* **2**: 69-73

Stein, J; Walsh, V. (1997). To see but not to read; the magnocellular theory of dyslexia. *Trends in Neurosciene* **20**: 147-151

Stein, JF; Riddell, PM; Fowler, MS. (1986). The Dunlop test and reading in primary school children. *British Journal of Ophthalmology* **70**: 317-320

Stein, JF; Riddell, PM; Fowler, MS. (1987). Fine binocular control in dyslexic children. *Eye* **1**: 433-438

Steinman, BA; Steinman, SB; Lehmkuhle, S. (1997). Transient visual attention is dominated by the magnocellular stream. *Vision Res* **37**: 17-23

Trick, LM; Enns, JT; Brodeur, DA. (1996). Life span changes in visual enumeration: The number discrimination task. *Developmental Psychology* **35**: 925-932

Index

T